W9-AMB-688

THE PEARLY GATES OF

CYBERSPACE

ALSO BY MARGARET WERTHEIM

Pythagoras' Trousers

THE PEARLY GATES
OF CYBERSPACE

A HISTORY OF SPACE

FROM DANTE TO

THE INTERNET

MARGARET WERTHEIM

W. W. NORTON & COMPANY

NEW YORK   LONDON

Lines from *The Divine Comedy* by Dante Alighieri, translated by C. H. Sisson, reprinted by kind permission of Laurence Pollinger Ltd. and Charles H. Sisson.

For information about permission to reproduce selections from this book, write to Permissions, W. W. Norton & Company, Inc., 500 Fifth Avenue, New York, NY 10110.

The text of this book is composed in Electra
with the display set in Base 12 Sans
Composition by RR Donnelley & Sons' Allentown Digital Services Division
Manufacturing by Courier Companies, Inc.
Book design by BTD/Mary A. Wirth

LIBRARY OF CONGRESS CATALOGING-IN-PUBLICATION DATA
Wertheim, Margaret.
     The pearly gates of cyberspace : a history of space from Dante to the Internet / Margaret Wertheim
             p.   cm.
     Includes bibliographical references and index.
     ISBN 0-393-04694-X
     1. Computers and civilization.   2. Cyberspace.   3. Internet (Computer network)
QA76.9.C66W48   1999
303.48'34—dc21   98-38200
                                                                              CIP

W. W. Norton & Company, Inc., 500 Fifth Avenue, New York, N.Y. 10110
            http://www.wwnorton.com

W. W. Norton & Company Ltd., 10 Coptic Street, London WC1A 1PU

1 2 3 4 5 6 7 8 9 0

For my mother
Barbara Wertheim
the space
of our conception.

CONTENTS

LIST OF ILLUSTRATIONS

PHOTO CREDITS

3.2 Alinari/Art Resource, NY

3.3 Alinari/Art Resource, NY

4.1 From *Unveiling the Edge of Time* by John Gribbin. Copyright © 1992 by John Gribbin. Reprinted by permission of Harmony Books, a division of Crown Publishers, Inc.

4.2 From *Hyperspace: A Scientific Odyssey Through Parallel Universes, Time Warps, and the 10th Dimension* by Michio Kaku. Copyright © 1994 by Oxford University Press, Inc. Used by permission of Oxford University Press, Inc.

4.3 From *Hyperspace: A Scientific Odyssey Through Parallel Universes, Time Warps, and the 10th Dimension* by Michio Kaku. Copyright © 1994 by Oxford University Press, Inc. Used by permission of Oxford University Press, Inc.

4.4 From *Hyperspace: A Scientific Odyssey Through Parallel Universes, Time Warps, and the 10th Dimension* by Michio Kaku. Copyright © 1994 by Oxford University Press, Inc. Used by permission of Oxford University Press, Inc.

5.1 From *Man the Square: A Higher Space Parable* by Claude Bragdon. 1912. Division of Rare and Manuscript Collections, Cornell University Library

5.2 From *A Primer of Higher Space* by Claude Bragdon. 1913. Science, Industry and Business Library. The New York Public Library. Astor, Lenox and Tilden Foundations

5.3 From *Projective Ornament* by Claude Bragdon. Dover Publications, Inc.

5.4 From *Projective Ornament* by Claude Bragdon. Dover Publications, Inc.

6.1 Photofest

7.1 Courtesy the author

7.2 © Circle of Fire Studios Inc. <www.activeworlds.com>

7.3 Copyright © 1997 Infobyte S.p.A.—Rome, Italy. All Rights Reserved

8.1 Alinari/Art Resource, NY

ACKNOWLEDGMENTS

I would like to thank my dear friend Howard Boyer, who signed this book to Norton, and who has believed in it all along.

No book reaches its final state without the input of readers who give generously of their time to wade through early drafts and make suggestions for improvement. In this respect I have been fortunate to have friends and family of high intellectual calibre. They are Brian Rotman, David Noble, Jeffrey Burton Russell, Alan Samson, Erik Davis, Cameron Allan, Barbara Wertheim, and above all my sister Christine Wertheim—the toughest critic a writer could have, but without whose insights this book would never have reached its current form.

I would also like to thank my editor, Angela von der Lippe; Neil Ryder Hoos for his invaluable work collecting the images; and Nan Ellin who suggested the title.

Finally, to my husband, Cameron Allan—who lived through the three-year creation of this work and made wonderful suggestions (and dinners) every step of the way—thank you for all your help.

THE PEARLY GATES OF

CYBERSPACE

THE PEARLY GATES
OF CYBERSPACE

Then I saw a new heaven and a new earth; for
the first heaven and the first earth had passed
away, and the sea was no more. And I saw the
Holy City, New Jerusalem, coming down out of
heaven from God . . . its radiance like a most
rare jewel, like jasper, clear as crystal. It had a
great high wall, with twelve gates. . . . And the
twelve gates were twelve pearls, each of the gates
a single pearl, and the street of the city was pure
gold, transparent as glass. . . . By its light shall the
nations walk; and the kings of the earth shall
bring their glory into it.
—THE BOOK OF REVELATION[1]

For the faithful Christian, death is not the end but the beginning. The beginning of a journey whose ultimate destination is
the Heavenly City of the New Jerusalem, the final Heaven,

wherein the elect will dwell forever in the light of the Lord. In this weightless city of "radiance," adorned with sapphire, emerald, topaz, chrysoprase, and amethyst, God himself "will wipe away every tear": "Neither shall there be mourning nor crying nor pain any more, for the former things have passed away." Along with liberation from pain, also will come the ultimate liberation, for "death shall be no more." There will be liberation also from internecine strife between nations. Here, people of all lands will walk together in harmony, while men pluck leaves from the Tree of Knowledge "for the healing of nations." "Nothing unclean" will enter this city whose pearly gates will always be open to those of pure heart.

The Heavenly City of the New Jerusalem was the great promise of early Christianity. An idealized polis, it is sometimes depicted in medieval imagery as a walled town floating on a bank of cloud (see Figure I.1). For those who adhered to the teachings of Christ, the Heavenly City was the final reward: an eternal resting place of peace and beauty and harmony, above and beyond the troubled material world. In the last centuries of the Roman era, as the empire disintegrated, such a vision offered special appeal. No matter the chaos and decay on earth, no matter that disharmony, injustice, and squalor reigned here, after death those who followed Jesus could look forward to an eternal haven of radiance and light. With that promise, Christianity was catapulted from just another obscure sect to the official religion of the empire.

So too, in our time of social and environmental disintegration—a time when our "empire" also appears to be disintegrating—today's proselytizers of cyberspace proffer their domain as an idealized realm "above" and "beyond" the problems of a troubled material world. Just like the early Christians, they too promise a "transcendent" haven of radiance and light, a utopian arena of equality, friendship, and virtue. Cyberspace is not a religious construct per se, but as I argue in this book, one way of understand-

FIGURE I.1. *Apocalypse of Angers* tapestry depicting the Heavenly City of the New Jerusalem coming down out of the skies.

ing this new digital domain is as an attempt to realize a technological substitute for the Christian space of Heaven.

Where early Christians conceived of Heaven as a realm in which their "souls" would be freed from the frailties and failings of the flesh, so today's champions of cyberspace hail their realm as a place where we will be freed from the limitations and embarrassments of physical embodiment—what cybernetic pioneer Marvin Minsky has derisively called "the bloody mess of organic matter."[2] Like Heaven, cyberspace supposedly washes us clean of the "sins" of the body, and, like Heaven, it too is being billed as a disembodied paradise for our "souls." "I have experienced soul-data through silicon," declared Kevin Kelly, executive editor of *Wired*, in a 1995 forum in *Harper's Magazine*. "You'll be surprised at the amount of soul-data we'll have in this new space."[3] "Our fascination with computers is . . . more deeply spiritual than utilitarian," writes cyberspace philosopher Michael Heim. In "our love affair" with these machines, he says, "we are searching for a home for the mind and heart."[4]

The notion of cyberspace as some kind of Heaven runs rife through the literature. In the influential collection of essays *Cyberspace: First Steps*, editor Michael Benedikt informs readers in his introductory remarks that "the impetus toward the Heavenly City remains. It is to be respected; indeed it can usefully flourish — in cyberspace."[5] According to Benedikt, cyberspace is the natural domain for the realization of a New Jerusalem, which he suggests "could come into existence only as a virtual reality." For Benedikt, "The image of the Heavenly City, in fact, is . . . a religious vision of cyberspace."[6]

In his essay, Benedikt mixes a yearning for prelapsarian innocence with a dream of postapocalyptic grace. He opines that "If only we could, we would wander the earth and never leave home; we would enjoy triumphs without risks, eat of the Tree and not be punished, consort daily with angels, enter heaven now and not die."[7] As he hints, cyberspace might make all this possible. Mark Pesce, cocreator of the virtual reality programming language VRML, has described one VR world as "a state of holy being which reminds us that, indeed, we are all angels."[8] Nicole Stenger, a virtual reality animator at the Human Interface Technology Laboratory at the University of Washington, offers the following testimonial: "On the other side of our data gloves we become creatures of colored light in motion, pulsing with golden particles. . . . We will all become angels, and for eternity!" She adds that "cyberspace will feel like Paradise."[9]

Robotics expert Hans Moravec, of the prestigious Carnegie Mellon University, imagines that in cyberspace we will find immortality, thereby realizing Revelation's promise that "death shall be no more." In his book *Mind Children*, Moravec writes ecstatically about the possibility of one day downloading our minds into computers so that we might transcend the flesh and live forever in the digital domain. He even envisages the possibility of resurrection. Here he imagines a vast computer simulation of our planet that would recreate in cyberspace the entire history of humanity. With such a simulation, he tells us, it should be possible to "resurrect all the past inhabitants of the earth," enabling everyone who ever lived to achieve immortality in cyberspace.[10] The book of Revelation promised the joys of eternity to virtuous Christians, but through the power of silicon, Moravec envisages it for us all.

No society's dreams take place in a vacuum. A culture's imaginings about the future, and its visions of what might be possible, or desirable, are always reflections of the time and of the particular society. What is it about *our* time and *our* society that is reflected in the "heavenly" appeal of cyberspace? In short, what does all this cyber-religious dreaming tell us about the state of America today?

As Umberto Eco and others have noted, America in the late twentieth century bears significant resemblances to the last years of the Roman Empire. In both places Eco points out that the disintegration of a strong centralized government and the collapse of the social polity leave each society open to internal rupture and fragmentation. Describing late antique Rome, Eco writes: "The collapse of the Great Pax (at once military, civil, social, and cultural) initiate[d] a period of economic crisis and power vacuum."[11] In this vacuum Eco locates the seeds of feudalism, where in lieu of central authority, powerful individuals emerged to lord it over disparate fiefdoms. In the twentieth century, the equivalent of these feudal lords are the multinational corporations, who likewise are filling the void of contemporary political "power vacuums."

As secular power dissipated in the ancient empire, more and more people turned to mystical, magical, and religious forms to provide new grounding and guidance in their lives. Like America today, late antique Rome was a cauldron of mystico-religious fermentation and all manner of sects flourished, from the ascetic number mysticism of Neoplatonism, to the hedonistic cult of Dionysus, and the Oriental cults of Mithra and Astarte. At the same time, a wave of religious fervor swept in from the Levant. Riding its crest were the followers of Jesus of Nazareth. With their enticing appeal of life everlasting in a heavenly paradise, and their promise of universal salvation, these "Christians" quickly gained adherents among the spiritually bereft Romans. At the end of the fourth century, under the emperor Theodosius, Christianity became the official state religion.

Sixteen hundred years later, Eco reminds us that "it is a commonplace of present-day historiography that we are living through a crisis of Pax Americana."[12] A rapid decline of central government and a fragmentation of empire are constant themes in the daily news. From the left, the homogeneity of the polity has been revealed as a dirty fiction now that women, minorities, and homosexuals are demanding to be heard; from the right, antigovernment sentiment explodes into open violence and rebellion. "Barbarians" too are pounding at our gates: the "Latin hordes" from the south who, we are told, would sponge off our Social Security and health care systems, and the "yellow hordes" of Asia, who are supposedly stealing our jobs with their cheap labor and undermining our economy with their crafty electronics and their mass-produced clothing.

Like the late Romans, we live in a time marked by inequity, corruption, and fragmentation. Ours too seems to be a society past its peak, one no longer sustained by a firm belief in itself and no longer sure of its purpose. As part of the response to this disintegration, Americans everywhere are looking to religion for new grounding in their lives. Whether it is the right-wing zeal of the Christian Coalition, California-style mysticism, or the pseudo–

Native Americanism of an executive retreat to a sweat lodge, U.S. society today vibrates with palpable spiritual yearning. Like the late Romans we too are searching for a renewed sense of meaning.

Cyberspace is not the product of any formal theological system, yet for many of its champions its appeal is decidedly religious. Not being an overtly religious construct is in fact a crucial point in its favor; for in this scientific age, overt expressions of traditional forms of religion make many people uncomfortable. The "spiritual" appeal of cyberspace lies in precisely this paradox: It is a repackaging of the old idea of Heaven but in a secular, technologically sanctioned format. The *perfect realm* awaits us, we are told, not behind the pearly gates, but beyond the network gateways, behind electronic doors labeled ".com," ".net," and ".edu."

The Christian Heaven, in many ways, was an extraordinary construct. Any vision with the power to endure for two thousand years would have to be. One of its chief features that resonated with the Romans (and with many others since) is that it has always been potentially open to everyone. People of all nations and skin colors can aspire to walk the streets of the Heavenly City. Unlike the Jews, from whom the early Christians were a breakaway sect, Jesus' followers opened their religion to all. One did not have to be *born* a Christian. There was no race or class requirement; a simple baptism would suffice.

Historian Gerda Lerner has noted that in this sense Christianity was a fundamentally democratic religion, perhaps the first in Western history.[13] One early mark of this democratic ideology was that during its formative years Christianity was especially welcoming to women. In Judaism, women were banished to a separate side of the temple, and the very covenant with God—the act of circumcision—was made only with males. Christianity posited no gender-specific covenant. In fact, religious historian Elaine

Pagels has shown that some early branches of Christianity, notably among the Gnostics, even allowed women to be priests.[14] One of the great innovations of Christianity was its promise of salvation for *all*, regardless of gender, race, or nationality. The kingdom of Heaven was open to anyone who embraced the teachings of Jesus.

Cyberspace too is potentially open to everyone: male and female, First World and Third, north and south, East and West. Just as the New Jerusalem is open to all who follow the way of Christ, so cyberspace is open to anyone who can afford a personal computer and a monthly Internet access fee. Increasingly, libraries and other community centers are also providing access for free. Along with the kingdom of Heaven, cyberspace extends its welcome to the whole human race—potentially so at least. Like the Heavenly City, cyberspace is also unfractured by national boundaries, a "space" where people of all nations can in theory mix together with mutual ease. Indeed, many cyber-enthusiasts would have us believe that the Net dissolves the very barriers of race and gender, elevating everybody equally to a disembodied digital stream.

As with Christianity, there *is* something potentially positive for women and racial minorities here, because the biasing baggage of a gendered and colored body is hidden from view behind the screen. Invisible and incarnate on the sea of cyberspace, here we cannot be summed up at a glance by the color of our skin or the bulges beneath our sweaters. The cybernaut is in some sense an idealized being, a creature of the ether. Thus at first glance we might agree with Pesce and Stenger that the cybernaut becomes a kind of angel.

One of the appeals of cyberspace is precisely the relief it provides from the relentless bodily scrutiny that has become a hallmark of life in contemporary America. In the bit stream, no one can see you wobble. Here, fat, wrinkles, gray hair, acne, baldness, shortness, and other aesthetic "sins" of the flesh are also hidden from view—all supposedly left behind in the "messy" material world. In this digital realm no one need know if you never work out or have an inordinate fondness for chocolate. Because online communication is primarily textual—at least for the moment—the cybernaut is freed from the constant pressure to look good. There is no such thing as a bad hair day in cyberspace, and here anyone who wants to can at least pretend to the illusion of perfection. Angels too were perfectly formed. Graceful, gorgeous, and adorned with that ultimate elegant accessory, wings, they were the medieval Christian beautiful people—a vision of what humans might be if only we could overcome the rough edges of incarnation.

In the bit streams of cyberspace one is also protected from the "evils" of *other* people's bodies. In the age of AIDS, cyberspace is a place to date in safety. During the seventies, Americans went to bars and discos in search of romance, in the eighties we went to gyms and clubs, but today such options carry heavy risks. Apart from catching a sexually transmitted disease, one could easily be raped or mugged on a night out in any major U.S. city. So much safer to go online, where romance can be found without ever leaving the living room. Tales of cyber-romance and cyber-sex are legion. What could be safer, or purer, than utterly disembodied love? A Christian ideal if ever there was one.

Some champions of cyberspace dream of escaping entirely from "the ballast of materiality," or what one commentator has called the "cloddishness" of the body. In *Neuromancer*, the prescient sci-fi novel that introduced the word "cyberspace" into our language, author William Gibson hailed "the bodiless exaltation of cyberspace."[15] Real-life virtual reality pioneer Jaron Lanier has said that "this technology has the promise of transcending the body."[16] Moravec too looks forward to a future in which the human mind will be "freed from bondage to a material body."[17]

There is, of course, nothing new about the desire to escape from the "messiness" and "cloddishness" of bodily incarnation. Western culture has carried that seed deep within it since at least the time of Plato, and in Christianity it has flowered in the Gnostic tradition.

Commentator Allucquere Rosanne Stone has noted that the apparent disembodiment engendered by cyberspace offers a special appeal to teenage boys.[18] With their bodies going through rapid and unnerving changes, Stone suggests that young males are drawn to a realm where they can escape their physical awkwardness and embarrassment. Unable to control their metamorphosing flesh, they flee to a place where they regain power and control. Ironically, young male cybernauts have coined a profoundly physical, even muscular metaphor for their travels in hyperreality. "Surfing the Net" does not conjure up images of disembodied ethereality, but a vision of physical power—both for the human user and for the space itself. Where the cybernaut is endowed with the prowess of a surfer, the space is likened to the ocean. Real-life ineptitude is replaced by an image of physical grace. It is no coincidence that hackers and surfers alike are predominantly young men—as also are the major Christian angels: Michael, Gabriel, and Raphael.

When the New Jerusalem arrives, one thing certain is that its citizens will not be lonely. Consider the vision of Heaven in the magnificent fresco by Giotto on the back wall of the Arena Chapel in Padua (see Figure 3.1). In this epic *Last Judgment* we witness a typical feature of medieval iconography. At the top of the fresco, behind Christ, stand rank upon rank of angels filling the heavenly Empyrean: The space is literally crammed full. Cyberspace too is teeming with people. There are one hundred million people already connected to the Internet and according to a recent Commerce Department report Net traffic is doubling every hundred days.[19] The collective nature of cyberspace is one of its primary appeals, as commentators continually stress. To quote Michael Heim: "Isolation persists as a major problem of contemporary urban society—I mean spiritual isolation, the kind that plagues individuals even on crowded city streets."[20] In the midst of this alienation, Heim says, "the computer network appears as a godsend in providing forums for people to gather in surprising personal proximity." The Net, supposedly, will fill the social vacuum in our lives, spinning silicon threads of connection across the globe.

Already cyberspace has become home to whole virtual communities, groups of people who meet and commune on the Net in chat rooms, on USENET groups, and in online forums. The San Francisco–based WELL community and the New York–based ECHO community are two of the more famous cyber-societies, with members of each group physically living all over the world. Cyber-giants such as CompuServe and America Online provide a veritable universe of meeting places and forums for their burgeoning cyber-citizenry. Chat rooms and USENET groups are available on almost any topic—from particle physics to neo-pagan goddess worship. Even those with the most arcane interests and obsessions can often find like-minded friends on the Net. As in the New Jerusalem, no one need ever be alone in cyberspace.

A further aspect of cyberspace that warrants our attention is the emphasis increasingly placed on image. Although at the moment most online activity is text-based, that is rapidly changing. With the advent of Web-based tools such as Java, many cyber-pundits believe the future is in pictures. Once every home is wired with fiber optics, instead of sending each other text messages we will be able to send real-time video messages. More intriguingly, I think, we will be able to send out into cyberspace graphically designed and animated "avatars" of ourselves to speak our words for us. Already, in the online cyber-city of AlphaWorld, visitors are represented by avatars that appear on the screen as cartoon-like figures walking through a simulated cityscape (see Figure I.2). Data also is being rendered into graphical form. Around the world researchers are working to reduce the ever-expanding reams of information into visually comprehensible terms.

Locking in on this ocean of images, politicians and cash-strapped schools are beginning to envision the virtual classroom. They imagine that the Internet will provide the inquiring mind with an endless array of interactive visual extravaganzas designed to keep even the dullest students interested. To quote former U.S. assistant secretary of education Dr. Diane Ravitch: "In this new world of pedagogical plenty, children and adults alike will be able to dial up a program on their home television to learn whatever they want to know, at their own convenience."[21] Online, Ravitch says, we will be able to dial up "the greatest authorities and teachers" on any subject, who will excite our curiosity and imaginations with wonderful videos and "dazzling graphs and illustrations." Why read when you could watch?

An emphasis on image was also a prominent feature of the Christian Middle Ages. In an age when illiteracy was the norm, religious images served to educate the populace about the Christian worldview. Paintings of biblical stories, of Christ, the Virgin, and the saints literally taught people about Christian history, cosmology, and morality. So too, we are now told that images will fill the educational vacuum in the age of cyberspace. In fact, some cyber-enthusiasts suggest that cyberspace is destined to become the very font of knowledge. As ever more libraries, databases, and information resources are made available online, the promise of god-like omniscience shimmers over the digital horizon. "Rightly perceived," says Heim, "the atmosphere of cyberspace carries the scent that once surrounded Wisdom."[22] As home to the Tree of Knowledge, the Heavenly City of the New Jerusalem also promised the fruit of ultimate Wisdom.

For better or worse, people are flooding into cyberspace. If

FIGURE I.2. In the cyber-city of AlphaWorld, visitors are depicted by animated "avatars" that can walk through the virtual streets and plazas of this online virtual world.

the current rate of growth were to continue, says MIT Media Lab director Nicholas Negroponte, "the total number of Internet users would exceed the population of the world" in the early years of the twenty-first century.[23] The tremendous embrace of cyberspace is a phenomena that cannot be explained merely by the availability of the technology. People do not adopt a technology simply because it is there. The basis of facsimile, for example, was patented in 1843, three decades before the invention of the telephone, but faxing did not take off as a widespread public tool until the 1970s — more than a century later. The Chinese invented the steam engine almost a thousand years ago, but they did not put it to use. And in the early 1980s the citizens of the U.K. showed no interest in Videotext (a pre-Internet technology that also operated over phone lines), even though it had been specifically developed for home use at a cost of many million pounds. As history repeatedly demonstrates, the mere availability of a technology is no guarantee that it will be taken up.

People will only adopt a technology if it resonates with perceived a need. For a technology to be successful, a latent desire must be there to be satisfied. The sheer scale of interest in cyberspace suggests there is not only an intense desire at work here, but also a profound psychosocial vacuum that many people are hoping the Internet might fill. The essence of this desire and the nature of this vacuum needs to be explained; we need to understand the factors that give rise to such intense interest in this particular technology. Specifically we might ask: What are the psychosocial conditions enabling cyberspace to become the focus of essentially religious dreams? What is it about our lives, and about cyberspace itself, that encourages such an outpouring of techno-religious dreaming?

L ike all human enterprises, cyberspace is embedded in a wide social matrix and any consideration of its appeal must look to broader cultural themes. In particular, as we contemplate the nature of cyberspace we inevitably run into the question posed by its enigmatic final syllable. What does it mean to talk about "space" at all? If cyberspace is a manifestation of this concept, then what exactly is it an instance of? The aim of this book is to look at cyberspace within the context of a cultural history of space in general.

At the heart of this story, as we shall see, is the age-old tension in Western culture between body and mind—in all its myriad manifestations, including that particular manifestation that Christians call "the soul." With respect to space, this tension has been played out in our shifting conceptions of what we perceive as *physical space* and *spiritual space*—that is, in our perceptions of a space in which our bodies are embedded, and a space in which our "psyches" or "souls" are embedded. It is within the context of Western culture's changing views about physical space and spiritual space that I hope to shed some light on the emerging arena of cyberspace and its powerful "religious" appeal.

It is one of the most remarked features of Western culture that for at least the past three thousand years our philosophies and religions have been deeply dualistic, splitting reality into a radical divide between matter and spirit. We inherit this dualism both from the ancient Greeks and from Judeo-Christian culture. For the Greeks, man was a creature of *soma* and *pneuma*, a body and a spirit. Pythagoras, Plato, and Aristotle all saw both human beings and the cosmos in bipolar terms. In the early Christian era the Greek *pneuma* was integrated into Judaic thinking and this amalgamation of Greek and Jewish intellectual currents gave rise to the theologically complex notion of the Christian soul.

During the thousand years of the Christian medieval era— roughly speaking from the fall of the Roman Empire in the fifth century to the start of the Renaissance in the fifteenth—Western intellectual culture was largely characterized by concerns pertaining to the soul. At least that is what medieval culture is primarily remembered for. Even this era's great physical achievements, such as its magnificent cathedrals, were religious projects, whose ultimate purpose was the enrichment of the Christian soul. But in the past half millennium—beginning in the Renaissance and more strongly since the "scientific revolution" of the seventeenth century—a profound shift has taken place, with Western attention increasingly turning away from the theological concept of soul and toward the physical concreteness of body. Since the eighteenth-century Enlightenment we have lived in a culture that has been overwhelmingly dominated by material rather than spiritual concerns. In short, in the modern West we live in a profoundly materialist and physicalist culture.

Unlike our medieval forebears, we modern Westerners have prided ourselves on, even defined ourselves by, our tremendous material achievements—our skyscrapers, freeways, and power stations; our automobiles, supersonic aircraft, space shuttles, and inter-planetary probes. In this modern physicalist age we have navigated the globe, put men on the moon, eradicated smallpox, worked out the structure of DNA, discovered subatomic particles, harnessed electricity, and invented the microchip. All these are extraordinary accomplishments, and ones we might well be proud of. In this sense also we are like the ancient Romans, for they too were a profoundly physical people, a culture that is remembered for its extraordinary feats of engineering and construction. Even today, a millennia and a half after the empire collapsed, visitors to the Latium countryside can still see long stretches of the aqueducts that carried water into Rome and they can still sit in vast stone amphitheaters that once seated thousands of Romans. The Coliseum and Pantheon endure as architectural wonders and Roman roads still crisscross Europe, some of them in continuous use for over two thousand years.

Modern mastery of the physical world is exhibited nowhere more strongly than in our scientific understanding of physical *space*. In the last five centuries we have mapped the whole of terrestrial space, as continents, ice caps, and even the ocean floors have yielded their secrets to our cartographers' skills. In the present century, we have also mapped the moon, and much of Venus and Mars as well. Our understanding of physical space now extends beyond our planet and out to the most distant reaches of the cosmos. After mapping the local solar system and detailing the relationships of the planets, astronomers have extended their gaze to the galaxies and mapped the structure of the cosmological whole. At the other extreme, particle physicists have been mapping subatomic space, probing the atom, then the nucleus, then the quark structures at very heart of matter. In this "age of science" we have mapped the physical universe at every level, from the vast scale of the galactic superclusters all the way down to the smallest particles. Moreover, neuroscientists are now mapping the space of our brains, probing inside our heads with PET and MRI scanners, gradually building up a sophisticated cartography of our gray matter.

Yet while we have been mapping and mastering physical space, we have lost sight of any kind of spiritual space. I do not mean to imply here that nobody in contemporary America or Europe has an inner spiritual life—clearly many do. I mean this statement in the very literal sense that we have lost any conception of a spiritual *space*—a part of reality in which spirits or souls might reside. In the modern scientific world picture it is a matter of cosmological fact that the *whole of reality* is taken up by physical space, and there is literally *no place* within this scheme for anything like a spirit or a soul to be. In the vision painted by modern science, the physical world is the totality of reality because within this vision physical space extends *infinitely* in all directions, taking up *all* available, and even conceivable, territory.

It was not always so. Where the modern scientific world picture recognizes only a physical realm, the medieval Christian world picture encompassed both a physical and a spiritual realm—it incorporated a space for body *and* a space for soul. This was a genuinely *dualistic* cosmology consisting of both a physical order and a spiritual order. A crucial element of this cosmology was that the two orders mirrored one another, and in both cases humanity was at the center. Physically, as in Figure I.3, the earth was at the center of the cosmos surrounded by the great celestial spheres that carried the sun, the moon, the planets and stars revolving around us. This was the old geocentric cosmology that prevailed from Aristotle to Copernicus. But more importantly, humanity was at the center of an invisible spiritual order.

In the medieval world picture the whole of the universe and everything in it was linked in a great spiritual hierarchy, some-

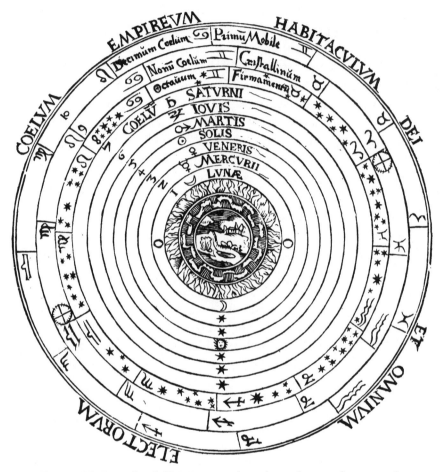

FIGURE I.3. In medieval Christian cosmology, the earth was at the center of the universe, surrounded by the concentric celestial spheres of the sun, moon, planets, and stars. "Beyond" the stars—and "outside" physical space—was the heavenly Empyrean of God.

times called the Great Chain of Being, that descended down from God. At the top of this chain, nearest God, were the ranks of angelic beings—the cherubim, seraphim, archangels, angels, and so on. After these "heavenly" beings came humans. After us were the animals, the plants, and finally the inanimate things such as minerals and stones. Within this scheme, man stood halfway between the ethereal beings of the heavens and the material things of the earth. According to medieval understanding, we were the only material creatures that also had an intellective soul, which latter property we shared with the angelic orders above. With one foot in both the physical and spiritual domains, we were the linchpin of the whole cosmic system: the halfway point and vital link between the earthly and heavenly domains. When medievals spoke of humanity being at the center of the universe, it was not so much our astronomical position they were referring to as our place at the center of this spiritual order.

Crucially, the medieval cosmos was *finite*—consisting of just ten celestial spheres centered on the earth. Beyond the final sphere of the stars was the very boundary of the physical universe, known as the Primum Mobile. Beyond this outermost sphere, and literally *outside* the universe, was the Empyrean Heaven of God. In truth the Empyrean was not only outside the universe, it was beyond space and time, both of which were said to end at the Primum Mobile. But *metaphorically*, medieval images of the cosmos depicted this heavenly domain beyond the stars, where there was, so to speak, plenty of "room" left.

What is critical here is that with the physical universe being finite one could imagine (even if, strictly speaking, only in a metaphorical sense) that there was still room available beyond physical space. Precisely because the medieval cosmos was limited in extent, this vision of reality could also accommodate *other kinds of space*. In particular it accommodated a vast region of "heavenly space" beyond the stars. Just what it meant to have a place *beyond* physical space is a question that greatly challenged medieval minds, but all the great philosophers of the age insisted on the reality of this immaterial, nonphysical domain.

In the scientific world picture, however, physical space came to occupy the *whole of reality*, leaving no room (even potentially) for any other kind of space to be. This vision, originally formulated in the seventeenth century, emerged out of a bold new *mechanistic* philosophy that envisaged the world not as a great spiritual hierarchy but as a vast machine. The consequences of this shift from the medieval to the mechanist world view continue to reverberate through Western culture and have transformed our conception not only of space, but also of *ourselves*. Tracing that shift is one of the primary themes of this book.

Chief among mechanism's founders was the French philosopher René Descartes, who is often portrayed as the arch-rationalist. Yet like all of mechanism's founders, Descartes was a deeply religious man who believed wholeheartedly in the Christian soul. In order to reconcile his mechanistic science with his belief in a soul Descartes made a radical philosophical move. He proposed that reality was separated into two distinct categories: the *res extensa*, or physically extended realm of matter in motion, and the *res cogitans*, an immaterial realm of thoughts, feelings, and spiritual experience. The purpose of the new science was to describe *only* the actions of material bodies in physical space, and so it applied only to the *res extensa*.

But if the *res cogitans*—Descartes' posited realm of mind and soul—was not to be encompassed by the new science, nonetheless he viewed it as a fundamental part of reality. His famous maxim, "I think, therefore I am," grounded reality not in the physical world but in the immaterial phenomena of thought. Thus, in its initial form, mechanism was a genuinely *dualistic* philosophy of nature: As with medieval thought, it took seriously the reality of both body *and* soul. Yet there was a fundamental, and ultimately fatal difference between the old medieval dualism and the new Cartesian version—one that would have profound consequences for "Man."

In the transition from the medieval to the mechanistic world picture a crucial shift occurred, for while the medieval cosmos was finite, the new mechanism suggested that the cosmos might be *infinite*. Once astronomers abandoned the idea of celestial spheres, there was ultimately no reason to suppose that the physical universe had any limit whatsoever. By the mid-eighteenth century that view had become scientific orthodoxy, and physical space was now seen to extend forever in all directions. But with physical space stretched to infinity, it was no longer possible to imagine (even metaphorically), that there was any room left "beyond" for any other kind of reality. In this new world picture there was in fact *no place left* for any kind of spiritual space to be.

One of the major effects of the scientific revolution was thus to write out of our vision of reality any conception of spiritual space, and along with that any concept of spirit or soul. This erasure precipitated in the Western world a psychological and philosophical crisis from which we have been struggling to recover ever since. Though Descartes would have been horrified, the end result of his mechanism was a desanctified and purely physicalist vision of reality. It is, therefore, a complete misnomer to call the modern scientific world picture *dualistic*, as is so often done. This world picture is entirely *monistic*, admitting the reality of the physical world alone.

Whatever Descartes' personal beliefs, mechanism set the West on a path that led rapidly to the annihilation of soul or spirit as categories of the real. The Enlightenment climate of the eighteenth century proved ripe for a hard-core materialism, and by the end of that century many agreed that the physical realm was the totality of the real. For men of the Middle Ages a world picture that encompassed only matter would have been inconceivable, yet almost inexorably that view came to prevail. In the new scientific world picture, humanity was no longer the linchpin of a cosmic spiritual hierarchy; we had become atomic machines. The old world picture with its striving souls, and its heavenly spiritual space, had given way to a purely mechanical universe in which the earth was just a lump of rock revolving in a vast Euclidian void. Moreover, while the medievals had seen humans as quintessentially spiritual and physical beings, the new mechanists regarded themselves in a purely physical sense. The monistic vision of space had thus been transformed into a monistic vision of *self*.

How did such a monumental shift occur? How did we go from seeing ourselves at the center of an angel-filled space suffused with divine presence and purpose to the modern scientific picture of a pointless physical void? What was at stake here was not simply the position of the earth in the planetary system, but the role of humanity in the cosmological whole. How did we go from seeing ourselves embedded in spaces of both body and soul, to seeing ourselves embedded in physical space alone? And critically, how has this shift in our vision of space affected our understanding of *who* and *what* we are as human beings?

What I aim to do in this book, as a prelude to a discussion of cyberspace, is to trace the history of how we in the West have seen ourselves embedded in a wider spatial scheme. I will follow the story of "space" from the late Middle Ages to today, specifically following the transition from a genuinely *dualistic* conception of physical space and spiritual space [Chapter One] to a purely *monistic*, purely physical view. We moderns are so used to thinking of "space" in geometrico-physical terms that it is hard for us to take seriously any other spatial system. Yet historian Max Jammer has stressed that "the use of a three-dimensional coordinate system . . . was not thought reasonable until the seventeenth century."[24] How did such a description of space become "reasonable"? As we shall see in Chapter Two, the answer has as much to do with the history of art as with the history of science; in particular it has to do with the rise of perspective painting and the increasing

[handwritten marginal note: this is always not the case, an anomaly]

Renaissance obsession with the realm of the body. In this respect, the new science would be prefigured by a new aesthetics.

Inspired by the new visual style, astronomers began to seek a new vision of the cosmos. During the seventeenth century, as "the mathematicians appropriated space" (to use the apt description of philosopher Henri Lefebvre), Western conceptions of both terrestrial space and celestial space underwent a revolution, giving rise eventually to the new Newtonian cosmology.[25] [Chapter Three] In this cosmology, celestial space was conceived not in terms of a heavenly spiritual order, but in terms of mundane physical forces and mathematical laws.

In our own century, the mathematical description of space has become a most complex business, leading first to a *relativistic* conception of space [Chapter Four], and finally to the bizarrely beautiful notion of *hyperspace.* [Chapter Five] With each new conceptual step, space has assumed an ever-greater role in the scientific vision of reality, until now it is seen by many physicists as the primary element of existence itself. Contemporary hyperspace physicists believe that in the end there is *nothing but space,* with even matter being just space curled up into minuscule patterns. In this vision, as we shall see, space becomes the totality of the real — the ultimate underlying "substance" of everything that is.

As we follow the history of space in its many manifestations, one of the themes we will be exploring in this book is how conceptions of space and conceptions of ourselves are inextricably entwined. For the medievals, who saw the world as an emanation from a divine spirit, it was impossible to imagine man without a spiritual dimension, yet for modern materialists who view the universe purely as a physical realm, man becomes *naturally* a purely physical being. Because we humans are intrinsically *embedded in space,* then logically we ourselves must reflect our conceptions of the wider spatial scheme. And so, as we trace the history of space, we will inevitably be looking at changing conceptions of ourselves.

In this sense space is revealed not merely as a "scientific" subject, but above all as a profoundly human one.

After tracing the history of pre-digital space, we turn, in the final portion of this book, to cyberspace. What sort of space is this new domain? How does *it* fit into the history of physical space and spiritual space that we have been considering? In fact, as we have seen, cyberspace itself is being presented as a new kind of spiritual space. If at first that may seem an odd move, I suggest that in the light of history, religious dreaming about cyberspace begins to make sense. Given the long history of Western dualism, a purely physical world picture was perhaps doomed to failure. As is now evident by the tremendous spiritual yearnings we see around us today, many people in the modern West—especially in America—are *not* content with a strictly materialist view. In this climate I suggest that the emergence of a new kind of *nonphysical space* was almost guaranteed to attract "spiritual" and even "heavenly" dreams.

No matter how often materialists insist that we humans are nothing but atoms and genes, there is clearly more to us than this. "I think, therefore I am," Descartes declared; and whether we modify "think" to "feel," or "suffer," or "love," what remains is the indissoluble "I," and deal with it we must. The failure of modern science to incorporate this immaterial "I"—this "self," this "mind," this "spirit," this "soul"—into its world picture is one of the premier pathologies of modern Western culture, and sadly, one reason why many people are now turning away from science. Sensing that something of fundamental importance has been occluded from the purely physicalist picture, they are looking elsewhere in the hope of locating this vital missing ingredient.

This omission is also a crucial factor in the appeal of cyberspace, for it is this immaterial "I" that cyberspace, in some sense, provides a "home" for. [Chapter Six] With this new digital space we have located an unexpected escape hatch from the physicalist dogma—for cyberspace too exists beyond physical space. Although it is true that cyberspace is realized through the by-products of physical science—the optic fibers, microchips, and telecommunications satellites that make the Internet possible are themselves all made possible by our tremendous understanding of the physical world—nonetheless, cyberspace *itself* is not located within the physicalist world picture. It is a fundamentally new kind of space that is not encompassed by any physics equations. As the complexity theorists would say, cyberspace is an *emergent* phenomena whose properties transcend the sum of its component parts. Like the medieval Empyrean, cyberspace is a "place" outside physical space.

There is a powerful sense in which, with cyberspace, we have manifested a kind of immaterial space of mind. When I "go" into cyberspace, my body remains at rest in my chair, but I—or at least some aspect of me—is transported to another "realm." Despite its immaterial nature, this realm is real. With more than one hundred million people spending time here, some of them for many hours a day, who can deny that cyberspace has become a genuine part of late twentieth-century reality (at least in the developed world)? As we shall see in Chapter Six, cyberspace is being specifically touted as a new realm for the "self." As MIT sociologist Sherry Turkle has written, "The Internet has become a significant social laboratory for experimenting with the constructions and reconstructions of self that characterize postmodern life.[26] Here, then, is a new space for the excluded Cartesian "I"—a technological *res cogitans*.

Moreover, as we shall take up in detail in Chapter Seven, cyberspace is also being presented as a new space of "spirit." With contemporary dreams of *cyber-immortality* and *cyber-resurrection* we have within the trappings of technology the reemergence of something very much like the old medieval Christian soul, something that I dub the "cyber-soul." I will argue that as we approach the twenty-first century we are witnessing the emergence of a new kind of dualism, a new version of the old belief that humans are creatures of *soma* and *pneuma*. In discussions about cyberspace and the fantasies surrounding it, we are seeing the reemergence of the old view that man is a bipolar being with a mortal material body and an immortal immaterial "essence," something that can potentially live on forever after we die.

This fusion of technology with religious ideals and dreams is not in fact a new phenomena. Science historian David Noble has shown that in the Christian West technology has been infused with religious dreams ever since the late Middle Ages. As he writes, when "artificial intelligence advocates wax eloquent about the possibilities of machine-based immortality and resurrection, and their disciples, the architects of virtual reality and cyberspace, exalt in their expectation of Godlike omnipresence and disembodied perfection" they are not doing anything "new or odd." On the contrary, this is a continuation of a thousand-year-old tradition.[27]

In particular, Noble has shown that in the Christian world technology has long been seen as a force for hastening the advent of the New Jerusalem. In his book *The Religion of Technology*, Noble traces the interweaving of the technical arts with the millenarian spirit and shows that from the twelfth century on, technology has been perceived as a tool for precipitating the promised time of perfection. On the eve of the scientific revolution, Johann Andreae, Tommaso Campanella, Francis Bacon, and Thomas More each envisioned a man-made New Jerusalem—a ficticious city in which technology would play a key role. Andreae's Christianopolis, Campanella's City of the Sun, Bacon's New Atlantis, and More's Utopia were all versions of idealized Christian communities notable for their use of technology. Today too, champions of cyberspace suggest that their technology will create a new utopia—a better, brighter, more "heavenly" world for all. With contemporary *cyber-utopianism*, the subject of Chapter Eight, the technology is digital rather than mechanical, but the dream remains the same.

And so at the end of our story the historical wheel comes full circle: back to dualism, back to "soul" (whatever that might mean in a digital context), and back to dreams of a New Jerusalem. What are we to make of such imaginings? How are we to interpret them in the light of our *own* times? These are questions we will explore in the closing chapters. Having examined these cyber-religious dreams I will, however, end this work on a note of my own optimism, one that seeks to interpret the potential of cyberspace not in a Christian utopian context, but in a context that opens out to the fantastic plethora of spaces which human cultures around the globe have conceived. For it seems to me that beyond the often naive rhetoric, cyberspace *does* offer us a powerful and potentially positive metaphor for how to understand the continuing enigma of its chimeric final syllable.

Before then, we begin our historical journey in the medieval era—a time when Europeans saw themselves embedded in both physical space and spiritual space. Our guide to this profoundly dualistic age is that supreme cartographer of Christian soul-space, Dante Alighieri.

SOUL-SPACE

H alfway along the journey of his life, the Florentine poet Dante
Alighieri set out on what has become the most famous journey
of the Middle Ages: a trip to the end of the universe and back.
Centuries before the advent of science fiction, Dante soared beyond the realm of the earth, past the moon and sun, on through
the planets, and out to the stars. He did not travel in a spaceship,
or any other kind of craft; his only navigational aid was the timeless wisdom of his guide, the Roman poet Virgil. That Dante was
accompanied by a man who had been dead for more than a thousand years signals immediately that we are not talking here about
any modern kind of space travel. Yet space travel is precisely what
the two poets were doing. Their journey, as depicted in *The Divine
Comedy*, is an epic elucidation of the medieval cosmos. As Dante
and Virgil travel from one pole of the universe to the other, we see
through their eyes a detailed geography of the entire medieval
spatial scheme.

Theirs is not only a journey through physical space (as in science fiction), but also through spiritual space, as conceived by the
Christian theology of the time. It is, above all, the voyage of a
Christian *soul*. Although Dante sets off on foot, seemingly in full
physical form, at the end of his tale he wonders whether he has

traveled in his body or out of it.[1] This uncertainty results from a key feature of the medieval world picture. In this dualistic scheme, body-space and soul-space mirror one another. In a very real sense Dante journeys both with and without his body. As an embodied being he travels the length and breadth of the material universe as understood by the science of his day; but simultaneously, he travels through the immaterial domain of soul, the realm that for the medieval Christian existed independently of body in the afterlife beyond the grave.

Here then was the starkest difference between the medieval and modern world pictures. Where our scientific picture encompasses only the body, and hence only the space of the living, the world picture of the Christian Middle Ages included the spaces of both the living *and* the dead. As a report to the living on the land of the dead, *The Divine Comedy* is the ultimate *map* of Christian soul-space. It is this space that we will be exploring in this chapter.

Yet if soul was paramount to the medieval mind-set, body was by no means irrelevant. Contrary to widespread misconception, Christians of the late Middle Ages considered the body crucial to human selfhood. So important, in fact, that the final stage of beatification in the soul's journey through the afterlife was signaled by its longed-for reunion with the body at the end of time — the resurrection of each individual person that was prefigured in Christ's resurrection from the grave. Only through unification of body and soul, said the great thirteenth-century theologian Thomas Aquinas, could man fully return to the state of grace in which he was conceived by the Creator of all things. Dante's poem takes us on a journey toward that beatified state.

Christian medieval soul-space was divided into three distinct regions or "kingdoms": Hell, Purgatory, and Heaven, documented successively in the three canticles of *The Divine Comedy*—the *Inferno*, the *Purgatorio*, and the *Paradiso*. As Dante depicts them,

Hell is a chasm inside the earth (Figure 1.1), Purgatory a mountain on the surface of the earth (Figure 1.2), and Heaven is coincident with the stars (Figure 1.3). After death, each soul would either be taken by a demon to the gates of Hell, or ferried by an angel to the shores of Purgatory, which Dante located on an island in the middle of the Southern Hemisphere. Only the *truly* virtuous—the saints and martyrs—were destined to go directly to Heaven; regular Christians must always expect some form of punishment after death. For them, the "second kingdom" of Purgatory functioned as a kind of preparatory school for Heaven.

Theologically, the middle kingdom of Purgatory stood between Heaven and earth, hence Dante represented it as a conical mountain, pointing upward toward God. In this middle kingdom, souls who were not sufficiently bad to be condemned to eternal damnation, but who had not led blameless lives, could work off the stain of their sin through the process of *purgation*—which entailed a series of cleansing torments. Yet despite these torments, souls in Purgatory were in a fundamentally different situation to those in Hell, because in Hell punishment was forever, whereas in Purgatory it was only temporary. In essence, Purgatory was "a Hell of limited duration."[2] Theologically speaking, souls in the second kingdom were on the same side of the ledger as those in Heaven, and *that* is where they too would ultimately go.

In *The Divine Comedy*, Dante journeys successively through each of these three kingdoms, leading us on a personal guided tour of the landscape of the medieval afterlife. Beginning at the gates of Hell, he first takes us spiraling down into the heart of darkness, ever deeper into the maw of sin. On coming through this horror-zone, we emerge at the foot of Mount Purgatory ready to begin the upward journey of salvation. During the trek up the Holy Mountain our souls are purged of sin, and thus cleansed we arrive at the mountain's peak, where the lightness of being engendered by a purified soul takes us effortlessly into the heavens.

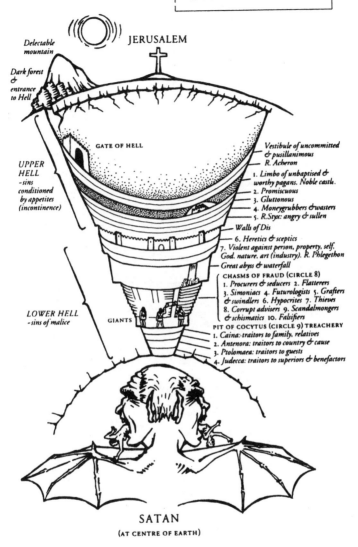

DANTE'S HELL

JERUSALEM

Delectable mountain

Dark forest & entrance to Hell

GATE OF HELL

Vestibule of uncommitted & pusillanimous
R. Acheron

UPPER HELL
- sins conditioned by appetites (incontinence)

1. Limbo of unbaptised & worthy pagans. Noble castle.
2. Promiscuous
3. Gluttonous
4. Moneygrubbers & wasters
5. R. Styx: angry & sullen

Walls of Dis

6. Heretics & sceptics
7. Violent against person, property, self, God, nature, art (industry). R. Phlegethon

Great abyss & waterfall

CHASMS OF FRAUD (CIRCLE 8)
1. Procurers & seducers 2. Flatterers 3. Simoniacs 4. Futurologists 5. Grafters & swindlers 6. Hypocrites 7. Thieves 8. Corrupt advisers 9. Scandalmongers & schismatics 10. Falsifiers

LOWER HELL
- sins of malice

GIANTS

PIT OF COCYTUS (CIRCLE 9) TREACHERY
1. Caina: traitors to family, relatives
2. Antenora: traitors to country & cause
3. Ptolomaea: traitors to guests
4. Judecca: traitors to superiors & benefactors

SATAN
(AT CENTRE OF EARTH)

FIGURE 1.1. Just as the medieval celestial spheres encode a metric of grace as one ascends up toward God, so the space of Dante's Hell encodes a metric of sin as one descends down toward Satan.

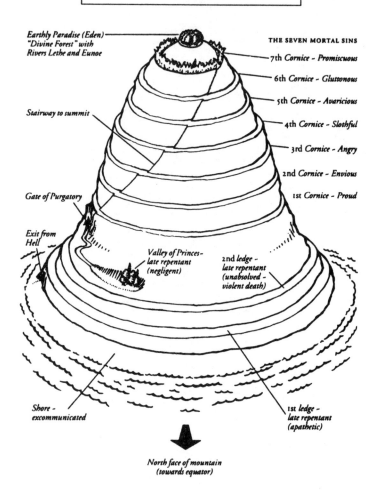

Earthly Paradise (Eden) —
"Divine Forest" with
Rivers Lethe and Eunoe

THE SEVEN MORTAL SINS

— 7th *Cornice - Promiscuous*

— 6th *Cornice - Gluttonous*

— 5th *Cornice - Avaricious*

Stairway to summit

— 4th *Cornice - Slothful*

— 3rd *Cornice - Angry*

— 2nd *Cornice - Envious*

Gate of Purgatory

1st *Cornice - Proud*

Exit from
Hell

Valley of Princes-
late repentant
(negligent)

2nd *ledge -
late repentant
(unabsolved -
violent death)*

Shore -
excommunicated

1st *ledge -
late repentant
(apathetic)*

North face of mountain
(towards equator)

FIGURE 1.2. For Dante, Purgatory is a conical mountain pointing upward to Heaven. Like Heaven and Hell, it too is organized as a spatial hierarchy.

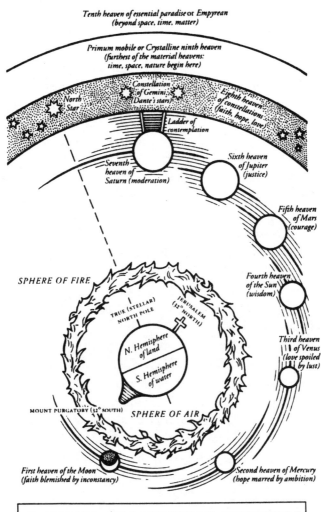

Tenth heaven of essential paradise or Empyrean
(beyond space, time, matter)

Primum mobile or Crystalline ninth heaven
(furthest of the material heavens:
time, space, nature begin here)

Constellation
of Gemini (Dante's stars)

North
Star

Eighth heaven
of constellations
(faith, hope, love)

Ladder of
contemplation

Seventh
heaven of
Saturn (moderation)

Sixth heaven
of Jupiter
(justice)

Fifth heaven
of Mars
(courage)

SPHERE OF FIRE

Fourth heaven
of the Sun
(wisdom)

TRUE (STELLAR)
NORTH POLE

JERUSALEM
(32° NORTH)

N. Hemisphere
of land

S. Hemisphere
of water

Third heaven
of Venus
(love spoiled
by lust)

MOUNT PURGATORY (32° SOUTH)

SPHERE OF AIR

First heaven of the Moon
(faith blemished by inconstancy)

Second heaven of Mercury
(hope marred by ambition)

DANTE'S GEOCENTRIC UNIVERSE
*showing also the virtues of the redeemed souls whom Dante
encounters on his journey in* Paradiso

FIGURE 1.3. In Dante's cosmos, the celestial heavens of the planets and stars
serve as a metaphor for the Christian Heaven—the realm of God and the
angels.

All this, Dante shows us in incomparable rhyming tercets. But if *The Divine Comedy* is first and foremost the archetypal journey of a Christian soul, it is also the story of a real historical man. Dante's genius was to weave together the Christian epic of "Man's" soul with the particular tale of his own unique life and times. Throughout, *The Divine Comedy* is peopled with real individuals whom Dante had known. As he travels through the afterlife, he converses with these souls, discussing the finer points of theology and philosophy, plus the intricacies of late thirteenth-century/early fourteenth-century Florentine politics. Even now, seven hundred years later, the partisans of Florence's bitterly warring political factions—the Guelfs and the Ghibellines—continue to regale us with their local squabbles. In this sense *The Divine Comedy* is a profound work of social commentary, a warts-and-all portrait of a fractious medieval community, at the center of which is Dante himself.

For Dante was not only a poet, but also—at least in the early part of his life—a deeply political animal. As a member of the Guelf faction, he seems to have thoroughly enjoyed the turbulent life of the Florentine political elite. Unfortunately, he got caught in the cross fire between the various factions and in 1302, while away on an ambassadorial mission to the papal court, he was tried in absentia by the opposing faction and sentenced to death. Unable to return to Florence, he never saw his beloved city again, and spent the rest of his life in exile.

Yet if exile was a bitter blow, it also turned out to be "a blessing in disguise,"[3] for it freed him to concentrate on his writing. No longer able to participate in politics, he embarked on the project of *The Divine Comedy*, determined to create nothing less than a new poetics—one that would weave together history, philosophy, and theology in an integrated whole. Written in vernacular Italian, rather than scholarly Latin, the poem is an extraordinary fusion of the secular and the divine, an audacious admixture unique in

Christian history. Dante himself seems to have regarded the poem as something like a new Gospel, and from the beginning that is how it was received. No other non-canonical Christian text has been so read, so analyzed, or so loved.

Having been banished from his home and friends, Dante created in *The Divine Comedy* a new life for himself. Denied a voice in Florence, he recreated himself in fiction and gave this poetic "self" a voice that would ring through the ages. What we have in the poem is, in effect, a "virtual Dante." In fact we know far more about this virtual Dante (what literary critics call "Dante-pilgrim") than we do about the real historical person ("Dante-poet"). It is this virtual self who speaks to us across the centuries and is our guide through the landscape of medieval soul-space.

As many commentators have noted, one of the great appeals of Dante's epic is that its world is so thrillingly real. Slogging through the fetid ditches of the Malebolge or trekking up the crisp terraces of Purgatory, you feel as if you are really there. You can almost smell the stench of the muck in Hell, hear the choraling of angels in heaven. This may be a journey of the soul, but few works of literature evoke the physical senses so powerfully. One hears, sees, smells the world Dante portrays. So real does this world seem that during the Renaissance there was a thriving tradition making intricate maps of Dante's Hell, complete with precise cartographic projections and measurements (see Figure 1.4).[4] Here, truly, was a rich "virtual world." As *The Divine Comedy* demonstrates so well, the creation of virtual worlds predates the development of contemporary "virtual reality" technology. From Homer to Asimov one of the functions of *all* great literature has indeed been to invoke believable "other" worlds. Operating purely on the power of words, books project us into utterly absorbing alternative realities. It is no coincidence the Bible begins with the phrase "In the beginning was the Word."

Yet *The Divine Comedy* is more than a work of literature,

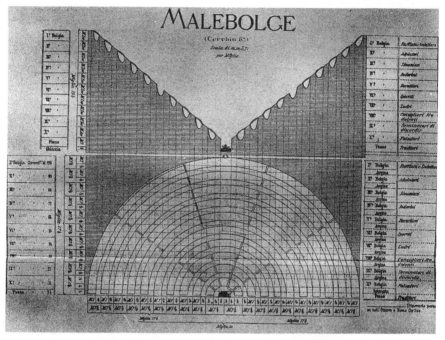

FIGURE 1.4. Cross section of the Malebolge from Agnelli's *Topo-cronografia* (1891). From the Renaissance on, people produced intricate maps of Dante's Hell, complete with detailed measurements.

and there is an important difference between the world Dante invokes and those of today's VR mavens. The crucial point is this: The "virtual worlds" being constructed on computers today usually bear little or no relationship to the world of our daily experience. For most VR pundits, *escape* from daily reality is precisely the point. Dante, however, was not trying to escape daily life; on the contrary he grounded his "virtual world" in real people, real events, and real history. Rather than trying to escape reality, he was obsessed with it. While it is true that in *The Divine Comedy* we find ourselves in a world populated by demons and angels, that we climb down the body of Satan and converse with the dead, we must remember that for Christians of the late Middle Ages all this *was* part of their reality. It was part of the grand metaphysical reality of which the physical world was just one small part. Rather than enticing us into an escape from reality, Dante invites us to see it whole, in all its vast *dualistic* scope.

Just as Dante grounded his epic in real human history, so also his realm of the afterlife is grounded in the physical cosmology and science of his time. His three kingdoms of soul-space beautifully parallel the general plan of the medieval physical universe. As in Figure I.3, that universe was geocentric, with the earth at the center surrounded by ten concentric "heavenly spheres," collectively carrying the sun, moon, planets, and stars around us. It is worth stressing that in this scheme the earth also was spherical. The notion of these sophisticated thinkers as flat-earthers is a myth, as historian Jeffrey Burton Russell has shown.[5] No serious scholar of the late Middle Ages believed the earth was flat, and indeed *The Divine Comedy* is full of references to the spherical shape of our globe. At the end of the Inferno, for example, Dante refers repeatedly to the southern "hemisphere."[6]

The basic plan of Dante's soul-space was that Heaven was coincident with the celestial realm, metaphorically surrounding and enveloping mankind in an ethereal embrace; Hell was inside the bowels of the earth, metaphorically speaking in the gutter of the universe; and Purgatory, as a mountain attached to the earth's surface, metaphorically pointed the way toward Heaven. All this was far from arbitrary; indeed the whole plan was governed by a rigorous logic internal to medieval cosmology and supported by the physical science of Dante's time.

As essential feature of medieval science and cosmology was the belief that the celestial domain of the planets and stars was *qualitatively* distinct from the terrestrial domain of man and the earth. On earth, everything was mortal and mutable, subject to death and decay, but according to medieval understanding the celestial realm was immutable and eternal. In the terrestrial realm everything was said to be composed of the four material elements—earth, air, fire, and water—but things in the celestial domain were supposedly made of the fifth essence, or quintessence, sometimes known as the "ether." The exact nature of this mysterious fifth essence was a source of much debate, but what is important here is that it was qualitatively different from anything in the terrestrial realm.

Medieval scholars believed that as one proceeded out from the earth, upward toward God, each celestial sphere became successively more pure and "ethereal" by virtue of its increasing proximity to the Supreme Being. From the earth to the Empyrean was thus a graduated scale of increasing purity and grace. Matter and spirit were in an inverse relationship, with pure matter (the earth) at the "bottom" of the universe, and pure spirit (God) at the "top." The whole cosmological scheme was like a great metaphysical onion, with the "lowliest" bit (the earth) at the core, and each consecutive layer gaining in perfection as one proceeded out and up. In effect, this universe encoded a metric of grace: The closer a place was to God, the more noble it was held to be, while the further away from Him, the less it was said to participate in divine grace.

Just as Heaven (the Empyrean), was at the *top* of the medieval cosmos, so, in the inherent logic of this system, the natural place for Hell was the rock *bottom*—that is, inside the earth as far away as possible from God. As the opposing spiritual pole to Heaven, Hell's location was inexorably determined by the logic of medieval cosmology. Purgatory, however, was a little more problematic. Because of the Middle Kingdom's association with sin, many authors located it underground, often inside a deep cave; but Dante chose a different (and rather imaginative) option. Befitting its status as the halfway house between Heaven and earth, he chose to envision Purgatory as a mountain thrusting upward toward grace.

For Christian medievals there was an ineluctable interweaving between the physical cosmos and the spiritual cosmos—the space of body and the space of soul. But since the spiritual realm was, for them, the *primary reality*, Christians of the Middle Ages oriented themselves first and foremost by a spiritual compass rather than a physical one. That this was so is evident from maps of Dante's time. Before the age of math-based cartography, European *mappa mundi* routinely depicted just a single landmass, the Northern Hemisphere, with Jerusalem in the middle. On these maps the Earthly Paradise (or Garden of Eden), was often drawn as an island off the far east coast, a detail gleaned from the Bible. For Dante and his contemporaries, the physical world was always and ever a reflection of the "true" underlying realm of soul, and it was into this primary reality that Dante would so memorably venture. Since his is a story of redemption, a journey up toward grace and light, naturally enough it begins at the bottom of the cosmos. Thus our exploration of soul-space begins with his, at the gates of Hell.

Above the entrance to the infernal kingdom Dante and Virgil are greeted by the famous warning popularly paraphrased as "Abandon hope, all ye who enter here."[7] For souls who cross this threshold, hope indeed becomes a thing of the past. Once a soul enters Hell its fate is sealed; it is condemned to punishment until the end of time and can dream neither of alleviation nor atonement. Ahead lies only torment and suffering for evermore. With the die thus cast, the human narrative ends. In Hell, there is literally *no future*. In Christian terms, the abandonment of hope is synonymous with the forfeit of redemption. By the magnitude of their sins, souls in Hell have thrown away the most basic Christian right, the salvation promised to all mankind by God's sacrifice of His son, Jesus Christ.

Dante's journey through Hell is a "descent" into sin, a downward spiral away from grace. The path he and Virgil follow is indeed a literal *spiral* that takes them winding down a long day's journey into night. Twenty-four hours is the precise length of time they spend in this metaphorical heart of darkness, a place where the sun never shines and where putrefaction reigns. Just as the medieval heavenly realm is structured in increasing levels of perfection as one proceeds upward through the celestial spheres toward God, so Dante's Hell is structured in decreasing levels of perfection as one proceeds down toward Satan. These are the famous nine "circles" of Hell. In essence, Dante's Hell is the infernal reflection, or negative, of his heavenly domain. Where the *external space* of the heavens encodes a hierarchy of grace, so, reciprocally, the *internal space* of Hell encodes a hierarchy of evil. As one descends into the Inferno, the magnitude and concentration of sin becomes ever greater, until at the bottom is Satan himself.

True Hell begins not in the first circle—a no-man's-land for the unbaptized and uncommitted, known as "Limbo"—but at the entrance to the second circle. It is here that every sinner must face the judgment of the monstrous Minos, the first of Dante's memorable cast of demons. As each soul approaches this ghastly creature,

> He sees what place in Hell is suited for it;
> And whips his tail around himself as many
> Times as the circles the sinner must go down.[8]

As soon as we pass by Minos' baleful glare we know immediately by the anguished cries that rend the air we have entered the infernal kingdom. And the deeper we descend the more dreadful will the wailing become.

In Dante's Hell, each circle is associated with a particular class of wickedness: In descending order they are lust, gluttony, greed, wrath and sloth, heresy, violence, fraud, and treachery. The deeper one goes the worse the sins are rated. (From our contemporary perspective, it is interesting to note that for the medievals lust was the *least* heinous crime.) As befits the general logic of medieval soul-space, punishments in the Inferno are suited to the crimes, and get more severe as we descend. Thus in the uppermost circle of the lustful, the punishment for illicit lovers is to be buffeted hither and yon by a bitter stinging wind. Their fate, so to speak, is to be endlessly blown by the uncontrollable winds of desire. By contrast, deep down in the eighth circle we find souls mired in boiling pitch, where they are mercilessly torn at by demons with hooks if they try to escape. As we drop deeper, both the torment and desperation increase. This is truly a descent into despair.

Spiraling down the abyss, each circle of Hell also gets progressively smaller as the sin becomes more concentrated. This increasing putrefaction of the soul is signaled in the environment itself, which becomes ever more dark, dank, and foul-smelling. More so even than the torments, it is the ambience of Dante's Hell that is so awful. One feels smothered by the inescapable rankness. The very space seems to be festering, and the sense of claustrophobia soon becomes unbearable. But contrary to the fire-and-brimstone image often associated with Hell, Dante's Inferno gets colder as one approaches the dark core where Satan resides. In the final circle, known as Cocytus, the souls of the treacherous are embedded in a lake of ice with only their heads poking out. Denied the possibility of motion, they cannot even try to run from their torment. The worst sinners don't have even their heads free; they are totally immersed in the ice—"like straws in glass"—condemned forever to freezing stasis.[9]

Yet again, this is in keeping with a rigorous internal logic. As we descend into the Inferno what we find is that souls are increasingly *confined* by their sins until those at the bottom, trapped in ice, are completely immobilized by the magnitude of their iniquity. Dante's message, poetically rendered, is that sin *imprisons*. And for no one more so than Satan. In the middle of Cocytus we find God's former right-hand angel, "the creature which had once been so handsome," buried up to his chest in ice.[10] A huge hairy giant with three monstrous faces, each mouth gnawing on a sinner, the beating of Satan's six great bat wings generates the chilling wind that keeps all Cocytus frozen. It is thus the evil one's *own* actions that keep him imprisoned. Here, at the heart of sin, we learn that *Hell is a place we make for ourselves*. And this is one of Dante's most powerful messages. By showing us how truly evil stifles, he hopes to help sway the reader back to the path of virtue.

If Dante's journey is first and foremost a spiritual journey, John Freccero alerts us that his descent into Hell may also be interpreted psychologically. "The inner space of Hell," he writes, "may be said to stand for the interior distance of a descent within the self."[11] For the late medievals the concept of "soul" encompassed not only those aspects of man that might relate to God, but also what we moderns would call the "emotions." In this pre-Freudian age the notion of a purely secular "psyche" was still half a millennium away, and the medieval discourse of "soul" ranged across a broad field that included many aspects of what we now know as "psychological" phenomena. Thus while Dante's journey is couched primarily in theological terms, it must also be seen as a metaphor for psychological transformation. Following the Augustinian injunction to "Descend, so that you may ascend," Dante also travels to the dark heart of *himself*.[12] Only after deep scrutiny of his own "inner life" can he reach "the zero-point" from which psychological healing can begin.[13] For Dante, that healing begins at the foot of Mount Purgatory.

Literary scholar Ronald R. MacDonald has argued that like the epic writers of Greece and Rome, Dante understood very well this psychological dimension of his text. Citing Dante, Virgil, and Aeschylus, MacDonald writes that "All these thinkers and poets teach in one way or another that through struggle and suffering and reflection, by submitting the self either individually or collectively to the worst as well as the best that lies buried within it, it is possible to effect a passage from a state of barbarity and disorder to a state of integration and harmony."[14] The journey out of Hell, and up the stairway of purgation to Heaven, must also therefore be seen as a kind of medieval psychotherapy.

Call it "purgation" or call it "therapy," the result is not just a purified soul but also a healed mind. In Freudian terms we could say that the journey out of Hell and up to Heaven represents the shedding of the ego, the letting go of that oh-so-heavy burden that *weighs* men down. For Dante, literally so—since his journey is at heart a quest for the perfect *lightness* of being. During the process of psychological healing enacted in *The Divine Comedy*, the inner space of mind is transformed from a hellish state of chaos and despair (metaphorically signified by the squalor of the Inferno) to a heavenly state of order and joy (signified by the blissful beauty of the Paradiso). Long before Freud, Christian theology encoded within it a sophisticated understanding of human psychology—as indeed do most religious and mythological systems.

For Dante, the process of psychological and spiritual transformation is enacted during the journey up Mount Purgatory, at whose base he and Virgil arrive after climbing back up to the surface from the bowels of Hell. After the stifling foulness of the Inferno, here in the Middle Kingdom his lungs fill with fresh air, the grass is green underfoot, and the sky shines blue overhead. The very environment vibrates with a palpable sense of optimism; one smells the scent of hope in the air. Here, as Dante tells us, "the human spirit cures itself, and becomes fit to leap up into Heaven."[15]

Because Purgatory was not explicitly mentioned in the Bible, the question of its exact location and nature was a source of much debate during the Middle Ages. Dante chose to locate it in the middle of a vast ocean in the Southern Hemisphere, directly opposite the globe from Jerusalem (note Figure 1.3). In the logic of *The Divine Comedy*, the line joining these two holy places defines an axis of salvation through the earth. By enduring torments on the Holy Mountain souls in the medieval afterlife atone for their sins, stripping away layers of wickedness in an ineluctable journey toward grace. As Jacques Le Goff notes, here "the ascent is twofold, spiritual as well as physical."[16] In *The Divine Comedy*, Mount Purgatory is, in effect, a medieval stairway to Heaven.

Where Hell was characterized by the death of hope, Purgatory could be defined as the place of hope. For the condemned, there is no exit from Hell, but "souls in Purgatory are on the move," constantly working their way up and out to Paradise above.[17] In opposition to the atemporal stasis of Hell, Purgatory is a place where time still has meaning. The process of purgation may be long and hard—one soul Dante speaks with has spent more than a thousand years there—but it is definitely a positive place.[18] Here the Christian narrative continues as the soul advances toward God. And from bottom to top the mountain resonates with hymns lifted in thanks to the Lord. Here, angels rather than demons guard each level.

As with Hell, Dante's Purgatory is also divided into nine distinct levels, known as "cornices," each more purified than the ones below. Again, the very structure of the space encodes the spiritual transformation being enacted, the "passage from a state of barbarity and disorder to a state of integration and harmony." The first level is the ante-Purgatory, where souls who repented late must serve out a period of waiting before being admitted to the mountain proper. This is the purgatorial equivalent of Limbo.[19] Moving into Purgatory proper, souls ascend through seven successive levels of *purgation*, or spiritual cleansing. Each level or "cornice" is associated with one of the seven venal sins, starting this time with the worst and moving up to the least heinous. In ascending order, they are pride, envy, wrath, sloth, avarice, gluttony, and lust. As in Hell, so in Purgatory punishments are fitted to the crimes. In the first cornice, for example, sinners carry stones on their backs, metaphorically atoning for the "burden" of pride. In the cornice of sloth, souls must counter their living lethargy with constant running, and in the cornice of gluttony the punishment is constant hunger.

But unlike Hell, Purgatory is not a nightmare. In contrast to the slime and filth of the Inferno, the Holy Mountain is carved into a series of crisp marble terraces, each adorned with elegant carvings depicting exemplars of virtue. Where the overall impression of Hell is messy and squalid, in Purgatory we find order and cleanliness. One immediately senses that here the war over chaos is being won. And where the path through Hell spirals *down* to the *left*—in Italian the word is *sinistre*—the spiral path around Mount Purgatory winds *up* to the *right*. Thus the very geometry of Dante's path through soul-space again encodes the moral meaning of his journey.

As a soul ascends up the Holy Mountain and the burden of sin is lifted, it becomes ever lighter. "In the Christian myth," Freccero notes, "it is sin rather than matter that weighs down the soul."[20] In other words, *sin is the gravity of soul-space*, the leadening force that pulls the soul away from its "true home" with God. With increasing lightness of being engendered by the process of purgation, the soul is drawn inexorably toward the heavenly Empyrean above. In Purgatory, then, the gravitational (downward) pull of sin is transmuted into "the levitational, 'God-ward' pull of sacred love."[21] After rising through all seven layers—thereby washing itself clean of each offense,—the soul emerges at the top of Mount Purgatory into the "Earthly Paradise"—the biblical Garden of Eden. As Dante scholar Jeffrey Schnapp explains, in Purgatory "the course of time is reversed, sin erased, the divine image re-stored."[22] Purgation thus unwinds the spiral of sin and takes us back to Eden, the cradle of our innocence.

The inescapably Christian context of Dante's journey is put into sharp relief by his and Virgil's arrival at top of the mountain. Here in the Earthly Paradise, Dante must leave behind his beloved guide, who for his part must now return to Limbo. According to medieval theology no one but a properly baptized Christian could enter into Heaven.[23] In the flowering woods of Eden, then, Virgil is replaced by a Christian guide, Dante's own personal "savior," the beautiful Beatrice. Object of perhaps the greatest-ever unrequited love story, Beatrice becomes here a universal symbol of Christian love. Again, however, the actual historical woman, Beatrice de Folco Portinari, is transformed into a *virtual* version of herself. And again, it is this virtual Beatrice we know today, far more so than the living woman, about whom we know almost nothing. With this heavenly lady as his guide, the virtual Dante, now metaphorically purged of his own sins, is "clear and ready to go up to the stars."[24]

To his astonishment, Dante finds that with the weight of sin lifted from his soul he is so light he rises effortlessly into the ce-lestial domain. Just as a river naturally moves down a mountain, so the virtual Beatrice explains that the unimpeded soul moves naturally up toward God.[25] Dante's journey through the celestial realm is not a trip to other physical "worlds," as in modern science fiction, but a kind of ecstatic cosmic dance through an increas-ingly abstract realm of light and motion. Here, luminescent choirs of angels fill the celestial space with heavenly harmonies—the mythical "music of the spheres."

Signaling that we have left behind the realm of flesh and pain, souls in Dante's Paradiso do not appear to him with their material forms—as they do in the first two kingdoms—here they are merely glowing forms of light. Moreover, following the

Neoplatonic association of light with grace, both the individual souls and the whole celestial environment become progressively more radiant. Light, as both fact and metaphor, is a distinguishing feature of the Paradiso. As also is motion. After the lugubrious plod through Hell and the slow climb up Mount Purgatory, the Paradiso puts the soul into warp speed. Dante and Beatrice zing through this heavenly space like "arrows" loosed from a bow.

As with the two lower kingdoms, this final region of Dante's soul-space is also organized into a ninefold hierarchy, this time melding naturally with the medieval hierarchy of celestial spheres. In the Paradiso we thus encounter an exquisite fusion of science and religion as Dante weaves together theological meaning and cosmological fact. Here, for example, the sphere of the moon is said to signify faith. But because the moon changes its appearance as it waxes and wanes, it becomes for Dante a symbol of faith blemished by inconstancy—as in the case of monks and nuns who deviate from their vows. Just as in the Inferno and the Purgatorio each level of the hierarchy was associated with a particular sin, so in the Paradiso each heavenly sphere is associated with one of the major Christian virtues: along with faith are hope, love, prudence, courage, justice, and moderation.

Yet if a hierarchy of sinners seemed justifiable in Hell, Dante is initially troubled by the celestial hierarchy of blessed souls. Surely, he suggests, *every* soul that is saved deserves to be as close as possible to God? Surely they should all be on the *same* level? In answer to Dante's queries Beatrice explains that each soul resides in the sphere that best matches its own spiritual nature. All are eternal, all are blessed, some just have a finer sensitivity to grace. This hierarchy is important for Dante, because the one feature his heavenly realm shares with his infernal realm is that both are spaces where time has effectively ended. As in Hell, souls in Paradise move neither up nor down the hierarchy; they are fixed forever in their spheres. Heaven, like Hell, is a dead end—a joyous and blissful dead end to be sure, but nonetheless a place where time has ceased.

Of all three regions of the afterlife, Heaven is the only one Dante has trouble describing. Where the *Inferno* and the *Purgatorio* each present a well-defined landscape and imagery, the *Paradiso* is famous for being so nebulous. In both lower kingdoms, the trials of the flesh provide the imagistic fuel, but the blissful state of the souls in the *Paradiso* offers few visual handles. As Dante and Beatrice make their ascent there are lots of joyous lights and great swathes of glowing mist, but there is no real geography. We are now in the realm of pure spirit, a space that, Dante admits, ultimately defies description. In the closing cantos of the *Paradiso*, when he at last enters the Empyrean, words finally fail Dante. The message—both concrete and metaphorical—is that in the presence of God we reach not only the limits of time and space, but also the limits of the language. Heaven might be the apotheosis of medieval soul-space, but precisely because of its *perfection* it is ultimately beyond human words. This is the realm of the ineffable.

The essential stasis of Heaven and Hell meant that the linchpin of medieval soul-space was really Purgatory. Only in the second kingdom did time continue to flow in a meaningful way. According to many medieval theologians purgatorial time was in fact the same as earthly time, the two spaces being bound together in the same temporal matrix. Moreover, medieval theology allowed that the purgatorial process could be affected by the actions of the *living*. In effect, the boundary between the land of the living and the second kingdom of the afterlife was surprisingly permeable. To quote Le Goff, Purgatory established "a solidarity . . . between the dead and the living," setting up a bond between the two worlds and serving as a convenient bridge between physical space and spiritual space.[26]

Because of Purgatory's "proximity" to the land of the living, it played a key role in medieval imagination. Starting in the early Middle Ages there was an increasing volume of literature relating to Purgatory, much of it describing visits by the living to the second kingdom, or visits by souls from there back to the earthly realm. With Purgatory, body-space and soul-space became, in effect, contiguous countries.

Prior to *The Divine Comedy*, the most famous purgatorial adventure was the medieval "best-seller" *Saint Patrick's Purgatory*, a twelfth-century tale that tells the supposedly true story of a trip through Purgatory by the knight Owein.

Like Dante, Owein enters the "other world" through a cavern in the ground, in this case one located in the grounds of a real church in County Donegal, Ireland. (The church is still standing today.) The infernal landscape of Owein's journey has more in common with the horrors of Dante's *Inferno* than with the clarifying vision of the *Purgatorio*, but the souls he encounters are also on their way to Heaven. In this version, Owein travels through fields where naked men and women are nailed to the ground and preyed upon by dragons and serpents. Elsewhere, they are boiled alive in molten metal or roasted on spits. Still others are hung by hooks through their eyes and genitals. All in all, it is a nightmare vision, but the journey ends happily in the Earthly Paradise, which Owein is told that he too can look forward to if he conducts his life properly.

Such tales clearly had a moral function; but they also served to feed medieval imaginations. Much as science fiction entertains us today with fantastic accounts of adventures in outer space, so Purgatory provided a setting for fantastic adventures in soul-space. Only a poetic genius like Dante could deal with the abstractions of Heaven, and only the truly audacious would dare to trespass in Hell, but Purgatory was a space in which the imagination and the narrative impulse could both run free. Owein was not the only one to venture into Purgatory through the County Donegal cavern.

This entrance to the afterlife had supposedly been shown to the original Saint Patrick by Christ himself, and ever since people had been making pilgrimages there to purge themselves of sin. The danger was that many who entered the hole apparently never came out.

In the imaginative ambience of late medieval Europe there was considerable traffic between the land of the living and the second kingdom of the afterlife. However, the majority of visitors to Purgatory did not try to go there bodily; they went purely in spirit, their souls leaving their bodies behind in a kind of medieval astral traveling. Such a trip was reported by the mother of the monk Guibert of Nogent, who witnessed there the trials of her deceased husband. In the other direction, the famed twelfth-century theologian Peter Damian recounted stories of a ghostly visit by a deceased man to his godson, and a visit by a dead woman to her living goddaughter. In the latter case, the ghost correctly predicted the goddaughter's demise. To the medieval mind, the boundary with Purgatory was highly porous: The dead made visits to the living and the living made visits to the dead. Body-space and soul-space were inextricably entwined.

For Christians of the late Middle Ages, the suffering of souls in Purgatory was very real indeed. Moreover, they believed the living had the power to *lessen* that suffering and help the dead more quickly through this time of trial. This could be done by offering up intercessional prayers, or by making special donations to the church. The practice was known as "suffrage," and during the late Middle Ages it came to play a key role in Christian life. Throughout Dante's journey through Purgatory, he is regaled by souls who beg him to remind their relatives back on earth to pray for them, and thereby lessen their torment.

Suffrage could apparently be very effective. In the eighth cornice of the *Purgatorio* Dante encounters the soul of his friend Forese who, although only dead five years, has already advanced almost to the top of the mountain, a feat he has accomplished because of the "devout prayers" of his loving wife Nella.[27] The responsibility for suffrages fell primarily on the shoulders of close relatives, especially spouses; but monks and nuns in religious orders also often "took prayer for the dead as one of their daily obligations."[28]

By binding together the living and the dead in a complex web of responsibility, suffrage created an "extension of communal ties into the other world."[29] The living on earth and the dead in Purgatory formed a kind of super-set of humanity, spanning what Le Goff slyly calls "the bogus boundary of death."[30] The living prayed for the dead not only out of charity for their dearly departed, but also because they hoped that when their *own* time came they in turn would be assisted by those they had left behind. Here was a kind of Christian Confucianism of the hereafter.

By extending the efficacy of Christian action into the afterlife, suffrage—which also included special Masses and church services—brought souls in Purgatory into the sphere of *clerical* power. "Even though God was nominally the sovereign judge in the other world," the Roman Catholic Church "argued it ought to have (partial) jurisdiction over them."[31] According to the influential thirteenth-century theologian Saint Bonaventure, popes even had the power to liberate souls entirely from purgatorial punishment. That is what Pope Boniface VIII saw fit to do in the jubilee year of 1300, when he decided to grant complete pardon of all sins to anyone who made a pilgrimage to Rome that year.

In practice, however, the theoretical power of popes to free souls from Purgatory was rarely used, for the Church was not so much interested in liberating dead souls as in maintaining a system whereby living souls would remain bound into the Christian web. It was in the clergy's interest that Purgatory should *not* be easy to escape, because the Church benefited mightily from suffrage payments for special Masses and other services. To put it bluntly,

"Purgatory brought the Church . . . considerable profit."[32] Not surprisingly, this system was open to abuse, and it led ultimately to a good deal of clerical corruption. Like border guards between any two nations, the clergy "patrolling" the boundary of body-space and soul-space all too often succumbed to illicit donations.

Abuse of the purgatorial system was especially egregious around the time of death. Even those who had led quite unsavory lives could, in theory, find salvation if they turned sincerely to God before their final breath. Within the system of suffrages, late repentance could also take financial form, and it was not unusual for wealthy men and women to make large donations to the Church in their closing years, or to leave such sums in their wills. It was these sorts of practices that led to the perception that some people were trying to buy their way into Heaven. That perception eventually caused Martin Luther and other Protestant reformers of the sixteenth century to condemn Purgatory as a Catholic abomination. Sadly, like justice systems the world over, the purgatorial system *was* a magnet for corruption; but rottenness in the ranks should not scupper the whole idea. Whatever the flaws in practice, in principle Purgatory was a gloriously humane invention, and one that has been sorely misrepresented in the modern age.

By providing a space for spiritual atonement Purgatory gave rise to what has been aptly called an "accountancy of the here-after."[33] Clearly some sins are more serious than others: What is the proper penance for each one? How much should purgatorial trial be reduced by penance done while the sinner is still alive? And how much by each suffrage offered after death? The exact time of a soul's stay in Purgatory would depend on the nature of the sins committed, the penance undertaken before death, and the intensity of the suffrages offered by the living after death. All in all this double-entry bookkeeping of the soul was a complex business.

A desire for a satisfying accountancy of sin was crucial to the emergence of Purgatory as a fully fledged kingdom of the afterlife. Because the Bible explicitly mentions only Heaven and Hell, it took a long time for the second kingdom to become properly established in Christian thinking. Not until the Council of Lyons in 1274 was it given formal theological countenance; before that soul-space officially consisted of just the two kingdoms. The coming into being of Purgatory is a rare instance in which we can see clearly the emergence of *a new space of being*. As such, there are important parallels with the creation of cyberspace today, and it is thus fascinating to see how this new medieval space emerged.

The idea of being held accountable for one's sins had, of course, always been central to Christian eschatology: As the Book of Revelation makes clear, no one escapes the Last Judgment. In Revelation, the author describes how he "saw the dead, great and small, standing before the throne . . . and the dead were judged by what was written in the books, by what they had done."[34] Accordingly, the good would be admitted into Heaven when the last trumpet sounded, no matter how lowly they had been on earth, while the wicked, no matter how high they stood on earth, would not.

But even divine justice is rarely a black-and-white matter. The strict polarity of the early Christian Heaven and Hell was at odds with the notion of a *merciful* God, one who by His very nature wants to accept as many of the faithful as possible into His eternal bosom. For example, asked the theologian William of Auvergne, what about the case of someone who is suddenly murdered?[35] Because such a soul would not have had the chance to atone for its sins before death, it would not qualify for admittance to Heaven, but as long as it hadn't committed any heinous crimes it hardly seems fair that it be condemned to eternal damnation in Hell. God's divine mercy in such cases was often signified in medieval literature by tales of angels fighting with demons for possession of newly liberated souls. Dante recounts such a struggle

over the soul of Buonconte da Montefeltro, a soldier killed in battle.[36]

Yet, the very concept of divine mercy also posed a dilemma,
because if God was simply going to forgive all sorts of sins immediately upon death, then what was the point in striving for saintly
behavior while alive? With too much divine mercy there would be
no *incentive* for saints and martyrs. In the long run, the Christian
concept of divine justice coupled with the notion of a merciful
God led almost inevitably to the need for a place in the afterlife
where souls who were not damnable could work off the taints of
their sins—a place that the *truly* virtuous would bypass. The early
Church recognized this need, but initially this spiritual cleansing
was said to occur as an instantaneous burning of the soul immediately after death. From the fifth century on, however, this instantaneous purgation was gradually transformed into the idea of
a place in which the corrupted soul would spend an extended period of time. Now, the greater a person's sins, the longer they
would suffer. Saints and martyrs would still go straight to Heaven;
but the rest would get a substantial period of punishment. With
Purgatory as the "middle kingdom" divine justice would be satisfied on all fronts.

The one troubling shadow over this rosy picture was the
problem of the "noble pagans," such as Dante's beloved guide
Virgil. How can it be, Dante asked, that people like Virgil who
lived before Christ was born are forever denied entry to the kingdom of Heaven? Moreover, how was it that people in Dante's own
time who lived in places like India and China, beyond the domain
of Christianity, were also damned? In short, how can those who
have never heard of Christ be held accountable for not having
been baptized into the "true" faith? The question is one the great
Italian epic never answers, and indeed "it haunts the poem."[37] Yet
if the late medieval concept of divine justice did not ultimately encompass *all* of humanity, for those within the Christian sphere

Purgatory extended the concept in a most beautiful way. As Dante well understood, Purgatory was the people's path to salvation.

Stories about journeys to and from the realm of the dead tend to evoke deep skepticism in we "scientifically-minded" moderns. The question thus arises: Whatever the exploits of the virtual Dante, did the actual historical Dante *really* believe in this vision of the afterlife? Did he and his contemporaries really believe there was a vast chasm inside the earth? Did they really believe in a terraced mountain opposite Jerusalem? Did they really believe in a set of heavenly crystal spheres? In a famous essay, Jorge Luis Borges has answered this question in the negative. That Dante believed in the "reality" of his vision is "absurd," says Borges.[38] But while Borges is right in one respect—Dante certainly never intended his poem in a *purely* literal fashion—I do not think the proposition is so "absurd." As Le Goff has written, journeys to the other world "were considered to be 'real' by the men of the Middle Ages, even if they depicted them as 'dreams.' "[39]

A major problem, I suggest, is that the very questions raised here are quintessentially modern. They are framed within the context of our purely *physicalist* paradigm, which was quite alien to the medieval mind-set. When we ask if Dante "really" believed in a set of heavenly spheres or a hellish chasm inside the earth, we are asking questions about *physical space*. In our minds we start wondering how far above the earth the lunar sphere would be. How far below the surface would the second circle of Hell be found? At what longitude might Purgatory be? We do this because we cannot help it. Our minds have been so trained—so brainwashed—to think of space in purely physical terms, it is almost impossible for us to think in any other way. It is not just that we have been to the moon and found no crystal spheres, or that we have circumnavigated the globe and found no terraced mountain; we simply cannot imagine a place being "real" unless it has a mathematically precise location in physical space.

From a purely physical perspective it is absurd to suggest that Hell is inside the earth or that Heaven is above the stars, but in the holistic scheme of Dante and his contemporaries these were the *logical* places for those realms to be. In the Christian medieval scheme, God was the organizing principle of space: His presence gave the universe an intrinsic direction, *up*, while sin created an intrinsic pull *down*. The internal logic of the system dictated that Heaven must be at the "top" of the universe and Hell must be at the "bottom." "Reality" could not be judged in purely physical terms, but must be seen in a broader sense that encompassed both physical and spiritual space.

There is another sense also in which Dante's world is "real"—the psychological sense. Just as Hell really is *within*, so too, psychologically speaking, Heaven is *out there*. It is no mere medieval foible that Dante's Heaven is beyond the earth, outside the self-obsessive chaos of sick men's minds and actions. To quote Le Goff: "As one moves from Hell to Purgatory and from Purgatory to Paradise the boundaries are pushed back, space expands."[40] Not just physical space, but also psychological space. *Room to move* is the essence of freedom for mind as well as body, and again a perfectly plausible psychological logic decrees that Heaven would be beyond the finite domain of the earth. If Hell is an inner cesspit where the psyche can barely "move," Heaven is an infinite field of rationation and love—a space that *should* transcend the finitude of our small material globe.

In Dante's cosmology, both soul *and* body are set free in the limitless space of Heaven. Ironically it is here, in that most ephemeral of all three kingdoms of the afterlife, where we find the two sides of man most inextricably entwined. It is in Heaven that the Christian body and soul at last become *one*. This heavenly integration serves not only as a final reminder of how deeply holistic was the medieval world picture, it also brings to the fore the paradoxes entailed in that holism. While Heaven is the apotheosis of the medieval spatial scheme, it is at the same time the most problematic. Precisely because it is God's domain, it is the place most difficult to reconcile with man.

The profound problems associated with the Christian medieval Heaven can be summed up in a single word: "resurrection." It is a commonplace of post-Renaissance propaganda that the medievals held the body in contempt, but in fact orthodox Christian theologians of the Middle Ages insisted that body was an essential component of human selfhood. Medieval theology held that at the end of time, when the last trumpet sounds, the blessed would be granted eternal life in body as well as soul. That was the promise they interpreted in Christ's resurrection and bodily ascent into Heaven.

In the Empyrean, the elect would sit in the presence of God whole in spirit, but also complete in flesh and blood and bone. The greatest of all medieval theologians, Thomas Aquinas, "explicitly said that soul separated from body is not a person."[41] For Aquinas, says historian Jeffrey Burton Russell, "Not only is the soul more human with its body than without it, it is actually more like God, because with the body its nature is more perfected."[42] Throughout *The Divine Comedy*, Dante echoes this theme, telling us again and again how the souls of the blessed long for the time when they will be reunited with their limbs.

Theologians of the twelfth and thirteenth centuries devoted considerable energy to discussions of just how the process of bodily resurrection would work. How would matter be reconstituted? How would separated parts, such as amputated limbs, be reconnected? Would fingernail parings be resurrected? Would hair clippings? Would circumcised foreskins? Would umbilical cords? But behind these questions lay a much greater dilemma: How can you have a body at all in a "place" that is, technically speaking, *beyond* space and time? Heaven—the true Heaven of the Empyrean—is only attained at the end of time, literally when the universe ends.

When the blessed finally go to Heaven to sit in the light of the Lord, like God they too will be in "eternity." Time and space will have ceased to be. The promise of "eternal salvation" does not mean salvation *for* all time, but rather salvation transcending time. Heaven is not *in* time; along with God it is *beyond* time. And also beyond space, for time implies motion and motion implies space. You cannot have one without the other. But if Heaven is beyond space, because space has ceased to be, then how can you have a body there?

As Russell writes: "It is not conceivable that creatures such as human beings with processes of senses, intellect and emotion could exist without space and time."[43] One cannot even sing a hymn without the existence of time, because "if there is no time, there can be no sequence."[44] Similar dilemmas apply to space because if there is no space there can be no extended bodies, hence no throats to sing. Some theologians of the late Middle Ages tried to get around these problems by considering Heaven the abode of "glorified bodies" rather than physical bodies, but as Russell wryly remarks, even for glorified bodies "it takes time to sing a hymn or to think a thought."[45]

The basic dilemma here is that "the idea of eternity works for God much better than it does for Heaven,"[46] a space that must after all contain human beings. Even if we cannot actually visualize a deity outside space and time, we can at least conceive of a transcendent divine being. But the notion of a transcendent *human* being is inherently problematic. Humanness, by its very nature, seems to be tied to both space and time. This is the puzzle that Dante confronts at the end of his poem. How can one envisage the heavenly Empyrean if it is a place beyond space? How can one imagine the souls enshrined there if there is, ultimately, no "there" for them to be? Dante's solution to this enigma is an ecstatic dissolution into geometry. Passing through "the skin of the universe" the virtual Dante looks out to see a Blazing Point of light around which circle nine rings of fire: God and the angelic orders symbolically rendered in light. Here, all directions and all dimensions fuse: "the Burning Point is not only the center, the innermost, but also the highest, the outermost" reference.[47] In this single point of infinite love is contained the whole of time and space.

No words can explain the "place" that is nowhere, the "point" that is everywhere. No metaphor can describe the fusion of body and soul into the Oneness that for medieval Christians was the source of everything. At the moment of this beatific vision, language at last fails one of its greatest exponents. Body-space and soul-space have been melded into one-space. The mystery is beyond intellection.

PHYSICAL SPACE

O n the front wall of the nave of the Arena Chapel in Padua is one of my favorite images in medieval art, and one that heralds a turning point in Western culture. The scene, which spans the top of the archway leading to the chapel's apse, is of the annunciation, that seminal encounter when God, through his herald the archangel Gabriel, makes Mary the mother of His son (Figure 2.1). On the left side of the arch kneels Gabriel and on the right side kneels Mary (Figure 2.2). Above this devout pair, spanning the space above the arch, flies a host of heavenly angels in celebration of the holy meeting below. Through the body and consent of Mary, God gives His son to humanity that we may be redeemed. The annunciation tableau has been painted thousands of times; it is one of the core images of the Christian canon, representing the supreme moment when the incarnation of the divine begins on earth. In the two kneeling figures, God and humanity—Heaven and earth—are conjoined.

But if the scene would have been wholly familiar to any early fourteenth-century visitor to the Arena Chapel, its rendition by Giotto was anything but that. Indeed, it must have been truly startling. Here was the medieval equivalent of *virtual reality*, images so compellingly solid and seemingly three-dimensional that view-

FIGURE 2.1. The Arena Chapel in Padua is a medieval virtual world.

FIGURE 2.2. Arena Chapel *Annunciation*. Like characters on a stage set, Gabriel and Mary appear to be in actual physical rooms set back behind the wall.

ers were meant to feel as if they were looking at actual physical figures in actual physical rooms. What is so arresting about the Arena annunciation is that Giotto has represented each of these figures in such a way that we seem to be looking *through* the wall into a "real" physical space *behind* the picture plane. It is as if the archangel and the Virgin are "really there" in a little virtual world of their own beyond the chapel wall.

In this startling rendition of Gabriel and Mary a revolution is heralded. We see here the first flickers of a new way of thinking that would eventually culminate in the modern "scientific" conception of *physical space*. The evolution and development of that vision of space is the story of this chapter—which, we shall see, weaves together the histories of both art and science.

With Giotto's Gabriel and Mary, we are immediately aware of a radical departure from the flat style of earlier medieval art. With Gothic imagery there had been almost no sense of depth or solidity, for those early artists were not interested in the illusion of three dimensions. In their images, figures floated against nebulous gold backgrounds, different parts of an image were painted at different scales, everything was flat and seemingly two-dimensional. In short, early medieval art was not "realistic." Giotto, on the other hand, was striving to simulate solid corporeal bodies occupying actual physical space. In his frescoes, buildings appear to recede into the distance; all objects are rendered at the same scale; and human figures seem to be made of solid material flesh. Moreover, his Gabriel and Mary are painted to look not just three-dimensional, but as if they have *weight*. Instead of floating airily, like Gothic figures, they seem anchored to the ground by a gravitational *force*. With this annunciation tableau we appear to be in the realm of regular *earthly physics*.

Giotto's simulation of physical space in his annunciation scene is further enhanced by faux architectural details that enclose the figures of Gabriel and Mary. On either side of the rooms in which the figures appear, Giotto has painted faux *sporti*, or small balconies. Where the rooms containing the figures appear to be set back behind the wall, the *sporti* appear to be jutting out from the wall into the physical space of the chapel itself. These faux features create a powerful illusion of actual architecture that has the effect of blurring the boundary between the virtual space of the image and the physical space of the chapel. With subtle illusionist artistry, they entice the viewer into a "virtual world" beyond the picture plane and suggest to us that it is "really" there.

The importance of these images in the history of Western art can hardly be underestimated. As John White has put it, "The frescoes painted by Giotto in the Arena Chapel in Padua, about the year 1305, mark an entirely new stage in the development of empirical perspective, as in every other aspect of pictorial art."[1] Now regarded as the founder of Renaissance painting, Giotto was the first artist to systematically explore the style that would eventually be codified as "perspective."

Yet this revolution in *representation* signaled far more than just the advent of a new artistic style. Underlying this move toward solid-looking images was a newfound interest in nature and the physical world—an interest that would eventually lead to the downfall of that grand dualistic vision so poetically articulated by Dante. In the long run, this new concern with the physical realm would constitute a major challenge to the medieval world picture, because the more people began to focus on the concrete realm of the body, the more they began to question the whole medieval vision of an ethereal spiritual realm. Ironically, at the very time Dante was immortalizing that vision, the seeds of its destruction were being sown.

No one in the early fourteenth century could have known the turn history would take. Certainly not Giotto or Dante, who, as French philosopher Julia Kristeva has remarked, "lived at a time when the die had not yet been cast."[2] Dante himself praised the new artistic "realism" in his *Divine Comedy,* and we know that he visited Giotto while the painter was working on the Arena Chapel. In the *Purgatorio,* the marble banks of the Holy Mountain are adorned with beautiful relief sculptures rendered in the new realist style, and the virtual Dante tells us these images are so lifelike that "Nature herself would there be put to shame."[3] So intent is he on examining these beautiful and convincing forms he can hardly bear to tear himself away. Which is quite how early visitors to the Arena Chapel might have felt, for here was an entire room filled with images painted to look as lifelike as possible. There had been almost nothing in the history of art to prepare a medieval visitor for such an experience. To Giotto's contemporaries the effect of the Arena Chapel must have been extraordinary: It was as if they had been projected *bodily* into the very life of Jesus Christ.

Here, one is engulfed from floor to ceiling in the world of Christ, his entire life story played out in startlingly naturalistic, three-dimensional, Technicolor splendor. The chapel is in fact a multigenerational homage to the Christian savior, for not only is each major event in his life depicted as an individual scene, so too we have the important events of his mother Mary's life, and of the life of her parents, Saints Anna and Joachim. As we see in Figure 2.1, this narrative is presented in three consecutive layers of imagery running around the walls of the chapel. To trace the story from start to finish, the visitor begins with the top row of images at the front of the chapel on the right-hand side. Following the story sequentially, one progresses along the right wall and then back down along the left wall. Having completed the top layer, you move down to the second layer, again progressing first along the right wall and then back down the left. Completing that, you move to the third and final layer. In a sweeping spiral one can thus follow the entire story of the Christian holy family.[4]

So "real" do these images seem with the new naturalistic techniques Giotto applied, you feel as if you could almost reach out and touch Jesus. Surely he is really "there," just beyond the chapel walls? And because it is a small chapel (built for the private use of the Scrovegni family), one has a sense of being cocooned in a little bubble universe. Rendered on plaster rather than in a computer, here nonetheless is a whole *virtual world*. Each scene in the cycle constitutes a separate virtual room, all linked together in an epic forty-part narrative. In the uppermost layer, the images recount the family's history *before* the advent of Christ: first the story of Anna and Joachim, then that of Mary herself as a young woman. Following this prelude, the middle layer begins the story of Jesus' life with the famous annunciation scene. From there we move on to his birth in the manger, the visit by the Three Wise Men, the young Jesus being baptized by John, the resurrection of Lazarus, and so on. The final layer depicts the story of the Passion, from Jesus' betrayal by Judas, through the Last Supper, the Crucifixion, the ascension into heaven, and finally the Pentecost.

But of course the viewer is not compelled to start at the beginning: He or she may dip in anywhere and follow one part of the story for a while before branching off to another. Today we would call it a *hypertext*. Almost eight hundred years before today's purveyors of computer-based VR, Giotto created in the Arena Chapel a *hyper-linked virtual reality*, complete with an interweaving cast of characters, multiple story lines, and branching options. In many ways this is a visual analog of Dante's *Divine Comedy*, a supreme medieval rendition of the Christian story in all its multilayered complexity. As we shall see, Giotto also painted here a grand image of medieval soul-space, complete with Heaven and Hell. Like *The Divine Comedy*, the Arena Chapel was intended to convey a comprehensive picture of the medieval Christian worldview.

It is widely acknowledged that Giotto is one of the great artistic geniuses of Western culture, but he must also be recognized as one of the great pioneers in the *technology* of visual representation. Although the frescoes in the Arena Chapel were not the first examples of the new artistic realism, White rightly notes that they were a quantum leap forward in the simulation of physical reality. Giotto understood, like no artist before him, how to simulate the effect of "being there." In doing so, he was responding not just as an artist but also as a "scientist." Although here we must remember that the distinction we now draw between "art" and "science" is a modern classification that was not so clearly made in the medieval era. More than any other representational style, the evolution of what came to be called "perspective" was driven as much by "scientific" as by aesthetic considerations. Above all, this new technological approach to image encoded Western man's newfound interest in nature and the physical world.

After a hiatus of some eight hundred years, the thirteenth century witnessed the return to Western Europe of natural science. In particular, the scientific works of Aristotle were reintroduced via the Arab and Byzantine worlds, and under the commanding influence of this ancient Greek polymath European scholars once again became interested in the workings of the physical world around them. Here was the start of the trend that would eventually give rise to modern science four hundred years later. In this vital, creative century, Petrus Peregrinus studied the properties of magnets and formulated the basic laws of magnetism; Robert Grosseteste studied the properties of light, spearheading the renaissance of geometric optics; and Albert Magnus studied plants, minerals, and stars. The astronomical works of Ptolemy and the mathematical works of Euclid both became the focus of intense study.

The new style of painting pioneered by Giotto and his contemporaries reflected this burgeoning empirical interest in the physical world. As White notes, these artists "were inquiring most acutely into what it was they could actually *see*, [they] were looking most intensely at the individual objects in the world around them and . . . trying to represent these objects more faithfully than their predecessors."[5] In the Arena Chapel, for instance, we see carefully observed images of sheep, goats, dogs, and plants (Figure 2.3). This sort of detailed naturalism was again a radical departure from the earlier Gothic style. Giotto's landscapes also are rendered with a new naturalistic sensibility. His mountains, while not in truth very "realistic," at least possess a plausible earthy solidity. His trees seem rooted in the soil, and his faces are portraits of distinct

FIGURE 2.3. Arena Chapel *Joachim's Dream*. Breaking away from the flat style of early medieval imagery, Giotto brought a new naturalism to painting.

individuals rather than symbolic representations of "man." Throughout, there is an attention to detail that was entirely new in Christian imagery. In short, the art of *empirical observation* was now being incorporated into the visual arts.

At school we are often taught that this move to a more naturalistic style represented a "maturing" of Western art. Just as the advent of modern science is said to signal our "progress" toward a "true" *understanding* of the world, so Renaissance art is often said to be a "true" *representation* of the world. But as art theorist Hubert Damisch has stressed, the new naturalism cannot be seen as some kind of Darwinian progress; rather it was a cultural *choice*.[6] According to Damisch, early Christian artists actively "refused [this choice] in a more or less deliberate and radical way."[7] These earlier artists did not paint in a flat iconic style out of ignorance, they had simply not been interested in portraying the concrete three-dimensional physical world; they were aiming for something quite different. Instead of representing the realm of nature and body, Gothic and Byzantine artists strived to evoke the Christian realm of the spirit.

In early medieval art, for example, Christ was often painted larger than angels, who in turn were painted larger than mortal men and women. Such disparity in size should not be interpreted as "childish" ineptitude; rather it was an attempt to signify the Christian spiritual hierarchy that we encountered in the previous chapter. The point is that early medieval art was not meant to be *representational*, rather it was *symbolic*. Perhaps the most graphic indication of this early artistic symbolism was the gold backgrounds characteristic of Byzantine art. Constructed from actual gold leaf, these backgrounds signified the presence of God, whose value was palpably evoked by the material itself. To quote philosopher Brian Rotman, gold, "intrinsically beautiful, changeless, precious, immutable, serves as the perfect icon" of God.[8]

Giotto, on the other hand, while still portraying religious subjects, was now deeply concerned with *physical verisimilitude*—with literal representation of physical phenomena. From the late thirteenth century, Western artists increasingly turned away from earlier symbolic styles and sought instead to represent concrete physical bodies in concrete physical settings. Thus in the Christ cycle of the Arena Chapel, all figures—Christ, angels, and mortals—are painted at the *same scale*. Here, the physical equality of bodies has replaced the medieval hierarchy of souls as the prevailing visual metric.

Absent also from Giotto's imagery is any hint of the gold background; instead, he attempts to represent environments naturally. His exterior scenes, for example, are marked by intense blue skies. While it is true that deep blue was sometimes used in Byzantine art to signify heavenly space, it also reflects the physical reality of the sky. (Full sky realism would be embraced in the fifteenth century with the addition of clouds.) But if Giotto's art undoubtedly still has links with earlier styles, it is also aiming clearly at a new physical literalism. The Christ cycle images in the Arena Chapel may be religious in content, but each scene has a solid material earthly setting. Figures here wear real human clothes, sit in real human chairs, and live in real human houses. The divine subject matter of the Christian canon is rendered into profoundly *human* terms.

Kristeva has noted that there is something deeply subversive about the Arena Chapel images, for in Giotto's tendency to naturalize and humanize he was literally *grounding* Christian imagery, wresting it away from its previous heavenly focus and bringing it down to earth. As Kristeva puts it, Giotto gave "a graphic reality to the 'natural' and 'human' tendencies of the ideology of his time."[9] In this sense his art reflects a profound shift in Western culture as Christian attention increasingly turned away from a "transcendent" realm of God and soul, toward the material realm of man and matter. With Giotto, Kristeva says, we witness "a subject liberating himself from the transcendental dominion."[10] In other words, attention was gradually shifting from what we have characterized as the domain of "spiritual space" to what would increasingly be understood as the realm of "physical space."

Another way of looking at this crucial transition has been pointed out by philosopher Christine Wertheim, who suggests that while early medieval artists painted what they "knew," Giotto and the new masters of the fourteenth century began to paint what they "saw."[11] In this sense, earlier medieval art must be understood as essentially *conceptual* (like a good deal of twentieth-century art). Gothic and Byzantine art had been attempting to convey an immaterial conceptual order, but the new naturalistic art of Giotto's time was specifically attempting to convey the *visual order* seen by the eye. As Wertheim explains, with the move toward naturalistic representation, the artist's "organ of sight" began to shift from the "inner eye" of the soul to the physical eye of the body. In other words, artists began to look *out* rather than *in*. This increasing privileging of the eye in our modes of representation has indeed been a unique feature of post-medieval Western culture—and, as we shall see, it would become a critical catalyst for the rise of modern science.

The most obvious example of this new visual trend in Giotto's work is his representation of buildings. Here we see most clearly the move toward what we now know as perspective. The faux *sporti* of the annunciation scene are but one example of the brilliant architectural illusionism that Giotto demonstrates in the Arena Chapel. Consider the marvelous effectiveness of Figure 2.4, *The Expulsion of the Merchants from the Temple*. Here Giotto has painted the temple from a slightly oblique angle so that we can clearly see two of its sides, both "properly" proportioned as they would appear in the physical world. Here also, doors and windows look as if they would really open; the columns of the portico look like they would really hold up the roof. True, the whole thing has a slightly cartoonish quality, and there are a few "off-beat" angles, but there is no doubt that we are looking at an image of a solid three-dimensional building. The *illusion* of solidity and depth, while certainly not perfect, is nonetheless convincing.

Throughout the fourteenth century the illusion of depth would become an increasing important priority for painters. Initially guided by intuition, the simulation of three-dimensionality

FIGURE 2.4. Arena Chapel, *The Expulsion of the Merchants from the Temple.*
Giotto's buildings begin to look genuinely three-dimensional.

would eventually be codified into a set of rules in the fifteenth
century. But long before the rule-based rigor of formal *linear per-*
spective, Giotto and the masters of the trecento honed their illusionistic skills, giving rise to a startling new realism that we today
recognize as the dawn of the Renaissance.

Yet if Giotto reveled in the physical world, he also remained
an artist profoundly concerned with the Christian realm of spirit.
At the same time that he portrayed the earthy physicality of bodies, he also painted angels illumined by an inner spiritual light
that, to my mind, is unequaled in Western art. If he was the first
artist to make the body look physically "real," he was also a mas-

ter of putting the Christian soul into the picture. It is this *dual* evocation which, I suggest, is a key to Giotto's enduring success as an artist. Unlike earlier medieval art, in which the figures are too impersonal to move our hearts today, Giotto's people are real individuals pulsing with joy and compassion and love. Here is the glory of bodily incarnation integrally imbued with a deep spiritual awareness. This synthesis is, I believe, a major reason why seven hundred years later Giotto's frescoes still speak to us with such force. He is the Dante of image and it is no coincidence the two men were contemporaries.

The age of Giotto and Dante—the early fourteenth century—was a time when Western culture was briefly poised between the two competing poles of spiritualism and physicalism. Where the early Middle Ages had been marked by a strong mistrust of the material world, the new naturalistic spirit of the twelfth and thirteenth centuries had reawakened European minds to the beauty, and glory, and sheer fascination of physical Creation. Both art and science blossomed under this influence. Yet, as we saw in the previous chapter, this was still an age of angels and demons, a time when Europeans still believed in the reality of an underlying spiritual realm. In this pivotal period, Giotto was striving to capture both a physical and a spiritual reality. Just as in verse Dante celebrated the journey of the Christian soul and the glory of the body, so in images Giotto reconciled the Christian person's dual nature.

There is no doubt, however, that the Western zeitgeist was changing. The shift away from the symbolic forms of Gothic and Byzantine art was also a move away from medieval theology's obsession with transcendence. It is not without reason that spiritual leaders through the ages have often viewed painting with suspicion (not just in the Christian West, but also in many other cultures). By seeking to *represent the world*, painting—especially realist painting—is a full frontal assault on the very idea of the *ineffable* that was the core of the medieval Christian vision of reality. (In its very vagueness, Gothic art had sought to imply this essential *unknowableness.*) Beautiful naturalistic images of the earthly realm threatened to divert attention away from the ineffable realm of spirit. Historically, this fear on the part of some medieval clerics would be fully justified, for as we now know that is ultimately what did happen. Yet from the beginning there were also clerics who saw the new naturalism itself as a boon to Christian faith — and here we encounter one of those crucial episodes in our story where theology would become a powerful spur to the evolution of thinking about space.

Foremost among this school of medieval thinkers was the English Franciscan monk and protoscientist Roger Bacon, who put forward a fascinating theological argument to justify the new artistic style. Bacon believed quite simply that realism in religious art could serve as a powerful propaganda tool for bringing unbelievers into the Christian fold. Indeed, art historian Samuel Edgerton has argued that Bacon's theological arguments provided a major impetus for the spread of the new realist style in Christian churches.

One of the more colorful characters of a truly inspired century, Bacon is sometimes referred to as the medieval Galileo. An early champion both of mathematics and experimentation, he spent his life promoting the cause of science and writing about its virtues. In the thirteenth century many theologians were resistant to the incursion of Greek science into Christian thinking, and Bacon appointed himself the chief defender against these naysayers. In 1267 he sent to Pope Clement IV a long treatise in which he outlined the potential value of science to Christendom. Here the new physicalist spirit was very much in evidence. According to Bacon, science would lead to all sorts of inventions that would improve the human condition. In his treatise he envisioned flying machines, automotive carriages, and machines for lifting heavy weights; also ever-burning lamps, explosive powders, a glass for concentrating sunlight to be used for burning enemy camps from afar, and magnifying lenses that would enable men to read small script at a great distance. In addition, Bacon said, science would lead to improvements in agriculture and medicine, and to elixirs for prolonging human life.

Yet Bacon's interest in science and mathematics was primarily focused on what they might do in the service of his faith. He had been motivated to compile his theories and send them to the pope after the failure of the Seventh Crusade to recapture Jerusalem in 1254. His aim in writing was indeed to inspire *another* crusade to drive the "infidel" out of the Holy Land, for he believed that science could be a key for reinvigorating Christian enthusiasm. In his treatise to Clement, Bacon extolled at length on the many ways in which science might serve Christian faith, but the one that concerns us here was its application to solid-looking imagery.

For Bacon, the key to the new realistic style of painting was the application of *geometry*. "Though he probably knew nothing of the relevant artistic activity in far-off Italy at the time," Bacon "was well aware of the power of visual communication, and became convinced that image makers . . . must learn geometry if they were ever going to infuse their spiritual images with enough literal verisimilitude."[12] In other words, Bacon believed that if artists understood geometry and applied it to their work they could make religious images look so *physically real* that viewers would believe they were *gazing at the actual events depicted*. According to Bacon, visual verisimilitude applied to subjects such as the life of Christ would convince people of the literal truth of the Christian stories, and thereby serve to convert them to Christianity. He called the new style "geometric figuring." By its power, he wrote, people would "rejoice in contemplating the spiritual and literal meaning of Scripture . . . which the *bodies themselves sensible to our eyes would exhibit* [my italics]."[13]

With almost preternatural foresight Bacon had perceived the psychological power of visual *simulation*. By the application of geometry to image, he tells us, bodies can become "sensible to our eyes." What we have here, seven centuries before the invention of the computer, is a clear understanding that "geometric figuring" could be the basis for an illusion so powerful that people would be convinced of the "reality" of what they were seeing. As the first person to comprehend the extraordinary illusionistic power of mathematically rendered images, Roger Bacon might justifiably be called the first champion of virtual reality. In particular, Bacon believed the new visual style could provide convincing simulations of biblical events—that it could, as it were, bring the Christian stories to *life*, and thereby serve in the battle against the hated Moslem "infidels."

By way of cultural comparison, history presents us here with a not insignificant irony, for the "infidels" had, in fact, their *own* sophisticated brand of "geometric figuring." Not perspective, but a highly evolved art of mosaic and tessellated pattern-making with which they adorned floors, ceilings, and walls. This beautiful Middle Eastern art form was itself the product of a culture richly imbued with mathematics. Yet this Arab art never sought to simulate physical reality; like Gothic art it aimed at a subtle symbolism in which a divine order was signified by the beauty of complex geometric patterns.

Now despite the fact that the Arab world had kept the flame of Greek mathematics and science alive for more than half a millennium—a service for which we in the West will be forever in their debt—Christians of the late Middle Ages unleashed on this sophisticated culture one of the most appalling spectacles in human history. For what other assessment can there be of the

Crusades? We cannot but mourn the opportunity lost where in place of the Crusades the West might have sought instead an alliance with the Moslem world. What wonders might have resulted if the two forms of "geometrical figuring" had been enabled to inform and enrich one another?

Just a decade after Bacon's treatise, ideas that he had advocated in theory were being put into practice. Consider a sequence of frescoes that predate even the Arena Chapel: the great cycle of images chronicling the life of Saint Francis in the Franciscan basilica at Assisi, the mother church of Bacon's order (see Figure 2.5). Dating to the last decade of the thirteenth century, this is the first known instance of a church filled with images consciously painted to look solid and three-dimensional—the first case of "geometric figuring" rendered on a grand scale.

FIGURE 2.5. The Basilica of Assisi—a virtual reality simulation of the life of Saint Francis.

Again the visitor to Assisi was meant to feel as if he or she had been projected into the world of the saint. Each major event in Francis' life was depicted as an individual scene that one could follow as a story around the church walls. Here was the beloved man giving his cloak to a beggar, here he was talking to the birds, and so on. Although we cannot be sure, many historians believe Giotto was also the master at Assisi. Whoever was responsible, these new lifelike images had an immediate impact. Saint Francis seemed to be positively leaping out of the walls, and "before the end of the thirteenth century" the Basilica of Assisi had "become the most visited shrine in all of Christian Europe."[14]

The Arena Chapel in Padua and the Basilica of Assisi are nothing less than technological marvels. Anyone who doubts these medieval churches warrant the title of "virtual realities" should ponder the fact that at Assisi the artists carefully painted faux architectural borders at the top and bottom of the images, and faux marble columns between them, with the express intent that these fake features should blend with the real architecture of the church. The physical space and the virtual space were thereby united. In later works by Giotto in the cathedral of Santa Croce, he even contrived that shadows in the frescoed scenes were painted as if illuminated by the actual physical windows. In all three churches, the virtual space of the images became an extension of the physical space of the building—another part of reality "beyond" the church wall. Even today when one visits these places, there is still an overwhelming sense of being transported to another "world." Contemporary VR craftsmen, with their billions of bits per second, might be able to conjure the illusion of *motion* (an impossible feat on plaster), but for sheer psychological force, today's practitioners of the digital arts could well learn something from the genius of Giotto.

Looking at Giotto's frescoes from our contemporary Cartesian vantage point, it is easy to imagine that he had a clear understanding of three-dimensional space. It is easy to imagine, in other words, that a modern understanding of physical space was already present in the late Middle Ages, and that artists simply had to develop the techniques for *representing* this space. Yet as Max Jammer has stressed, the idea of three-dimensional space was by no means clear in the fourteenth century.[15] Obvious though this particular view of space may seem to us today, it took a long time for such a conception to solidify in Western minds. For all the seeming modernity of Giotto's images, if we look closely we can see that he still reflects an essentially medieval understanding of space. Despite his cleverness in simulating depth, there are limits to his illusionistic power. And it is in these limits that we gain a fascinating insight into the huge psychological shift that Western minds would have to undergo before a truly "modern" conception of physical space could emerge.

Take a look at Figure 2.6, an image from the Assisi Basilica of *Saint Francis Banishing Devils from the City of Arezzo*. On one side of the image is a cathedral, on the other side is the city of Arezzo. Between these two architectural forms stands Francis commanding the demons away. Like a cloud of bats, these agents of Satan swoop upward and out of the city. Theologically it is a powerful image: the humble follower of Christ exorcising evil from a beleaguered town. Yet what concerns us here is not the religious message, but the buildings. Although each architectural block on its own is reasonably convincing, when we consider them together there is no unity between the two. Not only is each one painted at a different scale (the cathedral is almost as big as the entire city), they are portrayed from entirely different points of view. The cathedral is seen from the left-hand side, whereas the city is seen from the right. Each one is a separate disjointed element that seems to occupy its own independent space. In short, there is no sense of an *overall unified space*.

This sense of disjointed space is even more pronounced in

FIGURE 2.6. Basilica of Assisi, *Saint Francis Banishing Devils from the City of Arezzo*. Although the buildings here appear to be three-dimensional, each also seems to occupy its own separate space.

the image of Figure 2.7, *Saint Francis' Vision of the Celestial Thrones*. Again we notice that the altar at which the saint kneels is seen from a different point of view to the thrones. While the thrones are seen from the left, the altar is seen from the right. Again, each object is isolated in its own separate space. The point is that while the artists of Assisi could give the illusion of solidity to *individual objects*, they did not convey the idea of one unified physical space. In other words, these images have no *spatial integrity*.

Without such spatial integrity the illusion of physical reality is incomplete. That illusion would only be fully realized with the formalization of the rules of linear perspective in the fifteenth century. These rules (which in effect formalized Bacon's notion of "geometric figuring") gave artists a concrete recipe for representing all objects in the *same* three-dimensional space. More than anything, it is this spatial integrity that differentiates later images by artists such as Leonardo and Raphael. In these fully "Renaissance" images, everything appears not only at the same scale, but also from the same point of view. Most importantly, in later images all objects appear to occupy *one continuous, homogeneous, three-dimensional space*. It is precisely *this* conception of space that in the seventeenth century would become the foundation of the modern scientific world picture.

Long before the rise of modern science, painters played a crucial role in establishing this essentially geometric vision of space. As Edgerton has suggested, "geometric figuring" *retrained* the Western mind to see space in a new way. While Giotto in the fourteenth century did not have a clear conception of continuous Euclidian space, by focusing artistic attention on the simulation of *depth*, he and the other trecento masters set the West on a new course. Unconsciously, their new naturalistic artistic style helped to precipitate a revolution in thinking that would eventually demolish the great dualistic medieval cosmos, and would set Western humanity within a new spatial scheme.

FIGURE 2.7. Basilica of Assisi, *Saint Francis' Vision of the Celestial Thrones*. The thrones and the altar are each depicted from different perspectives. There is no overall spatial unity.

So conditioned are we moderns to think of space as a continuous *all-encompassing three-dimensional void* that it is difficult for us to imagine any other view. Yet it would be another three hundred years before that conception of space would be clearly articulated. To comprehend the massive shift entailed in formulating this new view of space, we must first understand how men of Giotto's time *did* see physical space. Like so much else about late medieval thoughts, their vision of space was inherited from Aristotle, and it is this Aristotelian view that we see reflected in Giotto's images.

Central to the Aristotelian conception of space was what is known as the *horror vacui*, or horror of the *void*—a belief expressed by Aristotle's famous dictum, "Nature abhors a vacuum." According to Aristotle, a volume of emptiness—what we would now call *empty space*—is not something that nature would allow. As the Greek philosopher Melissus put it, "the empty is nothing and that which is nothing cannot be."[16] Since Aristotle believed there could be no such thing as a volume of nothingness, he came to the conclusion that space itself could not have volume. Instead, he proposed that space is just the surrounding *surface* of objects. According to him, the "space" of a cup, for example, is just the ultra-thin surface where the cup meets the surrounding air. In Aristotle's conception of the world there are no empty volumes (no *voids*), because where one substance ends, another always begins. Consider a fish swimming in water. The water surrounds the fish completely, so where the fish ends the water begins. Likewise, where a cup ends, air begins. According to Aristotle there are no extended voids anywhere in the universe. In the Aristotelian world picture, matter fills every crevice, and space is just the set of boundaries that separates one material thing from another.

Strange though such a view of space may seem to us, it was founded on a deep belief in the *plenitude* of the universe. Stated simply, the Aristotelian universe is *full*. To Aristotle, the idea of a void was abhorrent, because that would imply a region of nothingness. Using his formidable intellectual powers he marshaled an impressive array of arguments to demonstrate that such a thing was logically *impossible*. Thus the very concept of space that seems so obvious to many of us today was considered by most scholars for fifteen hundred years to be actually impossible—even in principle. Moreover this Aristotelian abhorrence of the void translated neatly into the context of medieval Europe, for Christianity also had a theological tradition of an abundant Creation—a universe that God had created full.

In the Aristotelian conception, space has no volume, hence it also has no depth, being just the surface of things. From an Aristotelian viewpoint, only concrete *material objects* have depth—not space per se. This simple fact had profound implications for the new realist painters, because it implied that only individual objects could be painted with the illusion of depth, not the *intervening areas* between objects. That is indeed what we observe in the work of Giotto. In the image of Saint Francis at Arezzo, for example, only the buildings appear to have depth, while the space between them remains flat and Gothic. Likewise in the Arena Chapel, individual objects are convincingly three-dimensional but there is no sense of a three-dimensional space between things. In a sense, the objects themselves are Euclidian, but the surrounding space remains Aristotelian. By the time Giotto came to paint the Arena Chapel he was clearly becoming aware of this tension, and was looking for ways around it;[17] yet beneath the carefully constructed illusion of depth we can still discern an essentially Aristotelian vision of space. In this respect Giotto remained profoundly a man of the Middle Ages.

In the early fourteenth century, however, painters were not the only ones unconsciously striving toward a new conception of space. Scientific thinkers too were pushing at the boundaries of Aristotle's ideas. Despite the ancient logician's grip on late medieval thinking, there were those who rejected his views on space. In fact, from antiquity there had always been champions of void space, the earliest of whom predated even Aristotle. These were the ancient atomists, beginning with Leucippus in the fifth century B.C. According to Leucippus and his pupil Democritus, the material world was made up of indivisible particles, called "atoms," and between these atoms was void space. As readers will recognize, this basic configuration of atoms and void is in fact the view that would be adopted by modern scientists, and which is taught to us in school today. But before this *atomist* vision would be taken up in earnest, Aristotle's objections to void space had first to be overcome.

The beginning of a sustained critique of Aristotelian views about space dates to the late thirteenth century—around the same time that Roger Bacon was writing his treatise to Clement. This critique was in fact just one aspect of an important historical episode that is now recognized as one of the first serious clashes between science and Christianity. Yet again in this instance it was theology that would open the door to fruitful new ways of thinking about space. Central to this skirmish was the idea of "truth" and who had the power to determine it. For certain supporters of Aristotle, his ideas were so compelling they saw in him a new standard of truth—one to which they suggested even theology must be subordinate. Needless to say, theologians of a more orthodox bent were not amused at being told that Scripture should take a backseat to a Greek "heathen," and they fought back.

The points of contention were many, but in the story of space one would prove crucial. This was the contention that the universe is *immovable*. From an Aristotelian perspective, the immobility of the universe arose directly from the impossibility of void space. If one *was* to move the universe, that would leave an empty space behind; yet *that* was supposed to be impossible, so, *ergo*, it must be impossible to move the universe. From a Christian perspective, the implication was that not even God could do so. Traditionalist theologians were outraged by such imputed limitations to God's power, and they took action. Chief among the outraged was the bishop of Paris, Stephen Tempier, who in 1277 published a decree condemning 219 suspect philosophical views. The forty-ninth item on Tempier's list was the view that God was unable to move the universe on account that it implied existence of a void.[18]

Tempier's decree was vigorously opposed, and in 1325 it was finally revoked. Yet from the point of view of science the whole episode proved immensely fruitful. Paradoxically, this conservative theologian's objections to Aristotle had the effect of forcing scientific thinkers out of a rut. Most importantly, the furor over Tempier's decree precipitated a reexamination of Aristotle's ideas about space and motion—the upshot of which was that philosophers were forced to admit that the idea of void space was *not* a logical impossibility. Whether void space existed in *practice* remained to be seen, but from the end of the thirteenth century it had to be accepted as at least possible in principle. God *could*, in theory, move the universe. It is worth stressing here that it was in the interest of preserving a religious belief—the idea of an omnipotent God—that scholars were forced to rethink their scientific ideas about space. Contrary to contemporary dogma, religious ideas have often helped to spur the development of science—particularly the science of physics.[19]

After Tempier's decree the chains of Aristotelian thinking began to loosen, and in the fourteenth century there was an astonishing burst of creative scientific activity. The mere possibility of real void space opened up a whole range of questions that scholars eagerly explored. In particular, people began to consider the possibility of motion in a void. Thus in the early fourteenth century, as Dante was writing *The Divine Comedy* and Giotto was perfecting his painting techniques, we begin to see the emergence of a true empirical science of motion. A group of scholars in Paris known as the *terminists* and another group at Oxford, the *calculators*, defined such concepts as velocity and acceleration and began to formulate the basis for the modern science of dynamics. In short, by challenging Aristotle's views about space, these men of the Middle Ages began to pave the way for Galileo and the master physicists of the seventeenth-century.

The height of medieval thinking about space was realized by a brilliant Spanish Jew named Hasdai Crescas. As such, says Jammer, "Crescas made an outstanding contribution to the history of scientific thought."[20] It is perhaps not insignificant that it was a Jewish thinker who so advanced medieval thinking about space, for in Jewish mysticism there had been a long history of associating space with God. In Palestinian Judaism of the first century the word for place *(makom)* was also used as a word for God. From early on in Jewish theology the omnipresence of God was an important idea, and one that led eventually to the notion of space itself as an expression of God's ubiquity. As we shall see in the next chapter, the association of space with God would also prove enormously important in the thinking of Isaac Newton, the man who finally synthesized the modern scientific vision. Indeed, says Jammer, "a clearly recognizable and continuous religious tradition exerted a powerful influence on physical theories of space from the first through eighteenth century."[21]

In the early fifteenth century, Crescas become convinced of the reality of void space—not just in principle (as most of his contemporaries still believed), but also in practice. He came to this conclusion through a penetrating critique in which he demonstrated that Aristotle's own definitions about space led to logical absurdities. He pointed out, for instance, that under Aristotle problems arise when we try to talk about the earth's atmosphere. In Aristotelian terms, the "space" of the earth's atmosphere is its surrounding boundary with the first of the celestial spheres. But if that is so, Crescas said, what is the space of a small part of the atmosphere? Following Aristotle one would have to say that this is *also* the boundary with the first celestial sphere—in other words, it is the same as for the entire atmosphere. But that is clearly absurd. As Crescas noted, this is an endemic problem with the Aristotelian conception of space.

Consider the body in Figure 2.8a. Now look at 2.8b, which represents 2.8a with a piece cut out. Since Aristotle defines the "space" of a body as its containing *surface*, then the "space" of the part is *bigger* than that of the whole body. Clearly, this is absurd. Having pointed out such inconsistencies, Crescas went on to demolish all of Aristotle's objections to void space, and he convinced himself there were no intrinsic obstacles to its existence. According to Crescas, physical space was not the surrounding surface of things, but *the volume* that they occupied and in which they resided. More radically still, he championed the idea of an *infinite void* as the background to the whole universe. Unfortunately, Crescas was never able to bring his ideas to full fruition, for political instability in Spain during the fifteenth century put an end to the intellectual activities of Catalonian Jews.

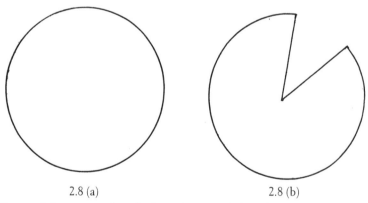

2.8 (a)   2.8 (b)

FIGURE 2.8. In Aristotelian thinking, the "space" of an object was equated with its exterior boundary. Thus, the "space" of a part was greater than the "space" of the whole.

Moreover, Crescas' ideas about space were far in advance of his time and would generally have to be rediscovered by others later. Despite his penetrating critiques, the bonds of Aristotelian thinking would only be broken slowly. To quote historian Edward Grant, while many late medieval scholars were prepared to speculate about the possibility of a *hypothetical* void, "the consequences of *real* empty space were too destructive of all that had come to represent the medieval world view."[22] If Aristotle was wrong about the void, then perhaps he might be wrong about other things as well—yet the whole medieval world picture was built around his science. As history shows, scientific thinkers did not succeed in overcoming Aristotle's ideas about space until the seventeenth century—and even then the idea of *infinite* void space would only win acceptance with the help of theological justification.

It is not often noted just how crucial a coherent conception of empty space was to the scientific revolution (and indeed to physical science ever since). Perhaps because this concept now seems so self-evident we tend to forget how difficult an idea it really is. That it is *not* self-evident is apparent from the enormous psychological difficulties it posed for most medieval and Renaissance thinkers. And logical argument alone was not sufficient to break down the deep-seated resistance. But as history would have it, a more powerful force than logic was at hand. It is here that we must return to the history of art and to the advent of formal *linear perspective*, for long before men of science accepted the new vision of space, it was *artists* who found a way to give coherent meaning to the idea of an extended physical void.

The key was their formalization of a set of rules for representing on a two-dimensional surface objects in three-dimensional space. Throughout the fifteenth century, painters such as Leon Battista Alberti, Piero della Francesca, and Leonardo da Vinci developed these rules. While these painters were not developing theories of space per se, but rather theories of *representation*, their pioneering work would prove crucial to the evolution of the modern concept of physical space that we know today.

Consider Figure 2.9, Piero della Francesca's *Flagellation of Christ*. Dating from the 1450s, a century and a half later than the Arena Chapel, we see immediately that a major transformation has taken place in the representation of space. On the left side of the image, Christ and his tormentors are seen in a room that now has all the hallmarks of three-dimensional Euclidian space. The lines on the ceiling and floor, and the row of columns at the right, all recede "properly" into the distance. Unlike Giotto buildings there are no "odd" angles here, and our minds accept the full physical

FIGURE 2.9. Piero della Francesca, *Flagellation of Christ*. One of the hallmarks of Renaissance imagery is spatial integrity. All objects appear to reside in one continuous, homogeneous, three-dimensional space.

viability of this room. Most importantly, all the elements are seen from the same point of view and they all occupy one *continuous space*. Unlike the spatial disjointedness that we remarked on earlier in Giotto's work, Piero gives us full *spatial integrity*.[23]

The illusion of spatial integrity, or spatial unity, which is a hallmark of the high Renaissance, is a major reason we interpret such images as the epitome of "realism." We feel on looking at a Piero or a Leonardo almost as if we are looking through a window to a scene beyond. And that was precisely what linear perspective aimed to simulate. In the first formal treatise on the subject, in 1435, Leon Battista Alberti explained the concept as follows:

> First of all, on the surface on which I am going to paint, I draw a rectangle of whatever size I want, which I regard as an open window through which the subject to be painted is to be seen.[24]

Alberti went on to elaborate a set of rules for conveying the illusion of seeing through "an open window," the method Renaissance artists would come to call the *construzione legittima* (the legitimate construction). Although in practice perspective is difficult to realize, the principle behind it is simple. Imagine, as in Figure 2.10, a canvas placed between the scene and the painter. This is our "window." Now imagine that a line, or "projecting ray," is drawn from each point in the scene to the painter's eye. As in Figure 2.11 the perspectival image arises from the intersection of all the individual projecting rays with the canvas. The image is, in effect, a *mathematical projection* of the three-dimensional scene onto a flat two-dimensional surface.

Alberti and his fellow Renaissance masters believed that with linear perspective they had found a way to simulate precisely what the physical eye sees. It is no coincidence their new style took its name from the old medieval science of vision and optics: *per-*

FIGURE 2.10. Albrecht Dürer woodcut. The aim of perspective was to simulate the illusion of looking at a scene as if through a window.

spectiva. For Piero and Leonardo, says historian Morris Kline, perspective was just "applied optics and geometry."[25] Their images were "real," they thought, because their rigorous geometric method of construction directly mimicked the human *visual field*. The transition that we began to see with Giotto was thereby completed. Physical vision had now supplanted "spiritual vision" as the representational ideal: The eye of the material body had replaced the "inner eye" of the Christian soul as the primary artistic "organ"

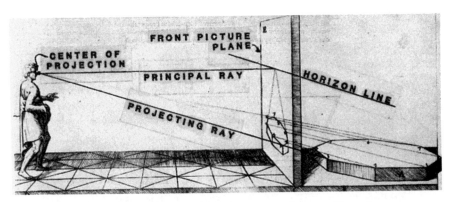

FIGURE 2.11. The perspective painter constructs an image as if seen from a particular point—the "center of projection."

of sight. Instead of an image being valued for its evocation of an invisible spiritual order, it was now valued by how closely the artist simulated the physical world. With this advancement in visual technology the spiritual symbolism of the Gothic period was swept away, and for the next five hundred years the framework of Western art would be overwhelmingly the space of the body.

The body. Renaissance art could almost be summarized as one long hymn to the human form. After the long era that the Renaissance would come to call "the Dark Ages," Western man had reawakened with a vengeance to himself as an incarnate being. Following in the footsteps of the ancient Greeks and Romans, Renaissance artists positively worshipped the human form. "Man" was the theme, not just in art, but also in literature and philosophy. The "ism" of the day was "human," as man in all his corporeal beauty again moved to the *center* of attention. Indeed what has become the very symbol of the Renaissance—Leonardo's famous drawing of "man" with his arms outstretched, inscribed in a circle and square—is a powerful comment on the idea of man as the measure of all things. From the fifteenth through the nineteenth centuries the body would rule supreme in Western art, reflecting the profoundly physicalist zeitgeist that has been the defining characteristic of the postmedieval age.

But even more important than perspectival imagery's *portrayal* of physical bodies was the fact that it also incorporated the *body of the viewer* into its spatial scheme. In many ways this was the most radical aspect of the High Renaissance style, and one that would prove enormously important in paving the way for the scientists of the seventeenth century. In this respect, painting would provide a model that both Galileo and Descartes would emulate. What is crucial here is not so much the virtual space of the image (that we have been discussing so far in this chapter), but the *relationship* of that image to the physical space of the viewer.

The key issue here is the disarmingly simple principle that a perspectival image is constructed from a single *point* of view—the so-called "center of projection." This unique point (seen in Figure 2.11) is not only the place from which the image is *constructed* (the point where the *artist* supposedly had his eye when he created the image), it is also the point from which the image is supposed to be *viewed*. It is the place where the viewer's eye also is meant to be. When looking at a perspectival image from the center of projection, as in Figure 2.12, the viewer literally "takes the place" of the artist: *His* or her eye replaces the artist's eye at the generative point of the scene. In effect, the perspectival image *directs* the viewer where to stand, because encoded in the image is the unique point in *physical space* from which the painting originates and from which it is supposed to be *received*.

What this means is that reception of a perspectival image is predicated on the "presence of a physically located, corporeal individual"—a concrete physical body in concrete physical space.[26] Because perspective encodes the position of the *viewing body*, it links together the virtual space of the image and the physical space of the viewer in a very formal way. The shift to perspective thus

FIGURE 2.12. Jan V. de Vries, *Perspective* (plate 30). A perspectival image encodes the "location" of the creating artist—and also of the viewing eye.

marked a shift not only in representation, but also in *reception* of images. Just as the physical eye of the body took over from "the inner eye" of the soul as the generative artistic organ, so by reverse, the physical eye also became the primary receptive organ. Unlike

Gothic art, which aimed directly at the Christian soul, perspective gives us images specifically for the eye. Again, this concrete placement of the viewing body by the perspectival image marks a radical shift from earlier medieval art. Gothic and Byzantine images could on the whole be viewed from *any* position. With their nebulous backgrounds and their lack of depth they had no particular "point" of view. As visions of the "inner eye" they encoded no relation to physical space and made no such demand on the viewer. Perspective, on the other hand, commands our reception from one particular physical point.

The rigorous linking by perspective of the physical space of the viewer and the virtual space of the image would enable later perspective artists to produce extraordinary feats of illusionism. A marvelous example is seen in Figure 2.13, the Baroque ceiling of the Church of Sant' Ignazio in Rome. The precise spot for the viewer to stand is marked on the floor by a disc of yellow marble. On looking up from this position the walls of the church appear to open out onto a dynamic skyscape where Saint Ignatius is being received into Heaven by an excited swarm of angels. It is almost impossible to tell where the real architecture ends and the faux begins.[27] Here, the subtle techniques of perspective make possible the illusion of a virtual reality which seems to blend seamlessly into the physical space of the church—one has the feeling of really "being there" beneath that angel-filled sky.

In terms of the history of space, the physically specific "point of view" encoded by linear perspective had the effect of making both the artist and the viewer of an image conscious of where they *themselves* were located in physical space. As a body of work, perspectival imagery thereby subjected Western minds to what amounts to an extended training course in conscious awareness of physical space. Depending on what point of view an artist chose to render, he could "direct" the viewer to stand *anywhere*. At first, artists like Piero and Alberti stuck to the simplest choice and

FIGURE 2.13. Fra Andrea Pozzo, *Saint Ignatius Being Received into Heaven.*
The ceiling of the Church of Sant' Ignazio in Rome appears to open out to an
angel-filled heaven above. A thrilling example of perspectival illusionism.

placed the point of view directly in front of the image, as if the viewer really was looking through "an open window." But the rules of perspective demand no such simplicity and soon artists were playing with quite bizarre points of view.

In Figure 2.14, Andrea Mantegna's *Saint James Led to Execution* (1453–57), the center of projection is actually below the *bottom* of the picture frame. If you want to look at this image from the technically correct "point of view" you have to focus on a spot on the wall *beneath* it. The effect is of a quite surreal image. Another example of an unusual perspective occurs in Leonardo's *Last Supper*, which has a center of projection fifteen feet above the floor. Only a viewer on top of a ladder could see this image from the technically correct position. Perceptual psychologist Michael Kubovy has argued that in contrast to earlier perspective artists, painters like Mantegna and Leonardo deliberately subverted the "open window" concept. In cases such as *The Last Supper*, the center of projection was consciously *separated* from the physical location of the viewer's eye. But there is more to these images than mere technical trickery. As Kubovy notes, these images amounted to a sophisticated form of mind game whose ultimate effect was, ironically, to bring back to perspective imagery a sense of detachment from the body.

Kubovy has shown that when we look at a perspectival image from any position *other* than the center of projection, our minds automatically adjust and we mentally see the image as if we *were* looking from that point.[28] It's as if the mind has a "virtual eye" that can roam around in space independent of the physical eye. Kubovy suggests that later Renaissance artists instinctively understood this, and that by separating this "virtual eye" from the physical eye they were specifically trying to induce in the viewer a kind of mystical disconnection from the body. Rotman has also shown that later artists such as Vermeer and Velázquez would take

FIGURE 2.14. Andrea Mantegna, *Saint James Led to Execution*. Perspective became the basis for subtle psychological tricks. Here the center of projection is located below the picture frame.

this separation to even greater extremes.[29] Thus, while perspective painting began by *embodying* a "point of view," it ultimately became once again a means for distancing the viewer from his or her body.

Just this process of removing the "point of view" from the physical eye of the body would also prove enormously important in the evolution of the modern scientific conception of space. By creating a virtual eye that was, in effect, free to roam about in space on its "own," this later phase of perspective provided people with a powerful psychological experience of *extended physical space as a thing in itself.* In effect, these Renaissance images set the mind free in a physical void, allowing people to "feel" for themselves this hitherto abhorrent concept. Without any conscious intention, perspective artists thereby succeeded in circumventing the strictures of Aristotle, and in a very powerful way they rendered the idea of extended void space real and palpable. In many ways this achievement is their most lasting legacy, for the free-floating "virtual eye" roving in space is precisely the model that Galileo and Descartes would adopt when formulating their new scientific world picture in the seventeenth century.

It would be going too far to credit perspective painters with the *discovery* of modern physical space. But as we have seen, "scientific" advances alone do *not* account for the huge psychological shift that had to take place before Western minds could accept this conception. To explain that shift I believe Edgerton is right when he says that without the revolution in *seeing space* wrought by the painters of the fourteenth through sixteenth centuries, we would not have had the revolution in *thinking about space* wrought by the physicists of the seventeenth.[30] Practically speaking, Giotto and his artistic descendants taught Europeans to look at space in a new way. No other culture before or since has taken the perspective experiment so far. In making the choice to go down that path, artists of the Renaissance unwittingly laid the perceptual and psychological foundations for a revolution in science.

The first person to clearly articulate the new vision of space in a scientific context was the pugnacious Italian who gave so much grief to the Jesuit priests, Galileo Galilei. Physicists today still refer to "Galilean space" when talking about the pre-relativistic variety and Einstein himself acknowledged a deep debt to his Italian predecessor. The notion of space that Galileo made the foundation of modern physics was just that depicted by the Renaissance painters: *a continuous, homogeneous, three-dimensional void.* Can it be a coincidence that this conception of space was adopted by scientists at the peak of perspective technique? Edgerton has noted that Galileo himself was well versed in the techniques of perspective and even applied for a position to teach perspective at a Florentine art academy.[31] For him, the "void" was no longer a matter for debate; it was the ontological grounding of reality itself, the neutral "arena" in which all things were contained and through which they moved.

With his razor-sharp reductionist mind, Galileo was able to abstract out of the world around him the seemingly essential features for a rigorous new physics. Above all, he proposed that an effective mathematically based science would require a mathematized version of space and time. In his new world picture "physical space" became at last synonymous with *Euclidian space,* a vast featureless three-dimensional void. At last, after two thousand years, Aristotle was defeated—the *void* had come to be seen as the very basis of existence. As with the ancient atomists, Galileo's universe consisted *only* of matter and void. For him, "the *real world* [was] a world of bodies moving in space and time."[32] Everything else—all the rich sensual qualities such as colors, smells, tastes, and sounds—were now to be regarded as just secondary, by-products of the "true" reality which was matter in motion in empty space.

Following the model of the Renaissance painters, Galileo abstracted the scientist's eye from his body and sent this virtual eye free-roving into space around him. Over the course of the next century this disembodied eye/mind would become the arbiter of the real. From now on the physicist's task was to seek out with his virtual eye the "essential"—i.e., the mathematically reducible— phenomena in the world around him. As with the perspective painters, the new physicists were seeking to represent in a rigorous mathematical fashion, physical relationships between material bodies in Euclidian space. For these scientists, Euclidian space was not just the background to reality, its very neutrality suppos- edly guaranteed that science itself would be neutral and objective.

It is difficult to overstate the magnitude of the philosophical shift which had taken place here. In Aristotle's vision of reality, space had been just a minor, and ultimately quite unimportant, category of existence, but in Galileo's vision this ephemeral entity was elevated to the *arena of reality*. As we have seen, hundreds of years were required for this vision to crystallize, and people had re- belled against it every step of the way. But now in the seventeenth century champions of the void were finally in the ascendancy. Their tabula rasa of emptiness was the slate on which these new physicists would boldly paint a new world picture.

The growing obsession with physical space spelled disaster for the old medieval world picture with its inherently spiritual spa- tial scheme. If the "real world" consists of material bodies moving in Euclidian space, *where* does that leave God? If the underlying substrate of reality is just an empty physical void, what *place* is there for the Christian soul? How indeed could humans, with our emotions and feelings and our longing for love, be accommodated in such an inherently sterile space? At the start of the seventeenth century the answer to these questions was not yet clear, but by the end of the century the whole edifice of medieval cosmology would have been swept away. The angel-filled spheres, the Great Chain of Being, the hierarchy of spirit, the purposeful strivings—all these would have been dumped like so much cultural garbage, and in their stead would be a new vision of the cosmological whole that for better or worse still dominates our lives today.

CHAPTER THREE

CELESTIAL SPACE

For Dante and Giotto, reality was intrinsically a twofold phenomena—as we have seen their universe consisted of both a
physical and a spiritual order. Moreover, in a complete inversion
of the modern materialist worldview, the late medievals regarded
the spiritual cosmos as the true or primary reality, with the physical cosmos serving as an allegory of this ultimate domain. Within
this philosophical framework, says Jeffrey Burton Russell, natural
science was "an inferior truth pointing to the greater truth, which
[was] theological, moral, and even divine."[1] The primary concern
of the medieval artists and philosophers was the "ultimate reality"
of the spiritual cosmos, which for them was "God's utterance or
song."

This *other* reality also was represented in the Arena Chapel,
where the entire wall facing the altar is given over to an epic depiction of medieval Christian soul-space. Here, as in Figure 3.1,
we find Giotto's monumental image of the Last Judgment. Just as
Michelangelo would do two centuries later on the back wall of the
Sistine Chapel, Giotto devoted the pride of place in his Paduan
chapel to the Christian cosmology of soul. It is immediately apparent that this image is in stark contrast to the Christ cycle of images we considered in the previous chapter. Signifying that we

FIGURE 3.1. Arena Chapel, *The Last Judgment.* For Giotto, spiritual space could not be rendered according to the dictates of naturalistic illusionism.

have left behind the physical realm, there is almost no attempt here at the illusion of three dimensions, for the Christian soul is not to be bound by the laws of Euclidian geometry. Nor will it be bound by terrestrial physics, which Giotto also now leaves behind. Instead of figures being anchored to the ground, as they are in the Christ cycle, here in the spirit world they float against an intense blue background—a depthless field signifying *divine space*.

A hierarchy of character is also immediately evident in this image, with the central figure of Christ now a monumental presence dwarfing all other personae, and the angels and apostles in Heaven occupying the next rung down the scale. Below them, on the left-hand side of the fresco, are the ranks of the saved. Fulfilling the promise of resurrection, they emerge from their graves as tiny figures, gaining size and stature as they ascend into Heaven above. On the right side of the painting lies Hell, safely walled off behind scarifying rivers of fire. Appropriate to their puny spiritual stature, figures here are minute; even Satan, lord of this infernal kingdom, is significantly smaller than his spiritual counterpart, Christ.

This powerful image reminds us that for Giotto and his contemporaries, the world could not be reduced solely to *physics*. Glorious though it may have been to render bodies and buildings on earth in geometrically correct proportions, Europeans of the fourteenth century never lost sight of the spiritual dimension of their profoundly Christian reality. In particular, they believed that "beyond" the physical realm of body (so gracefully celebrated in the Christ cycle), there was the eternal mystery of Heaven. That Heaven is a fundamentally *different* plane of reality is explicitly signified by Giotto in *The Last Judgement*.

Notice in Figure 3.1 the two angels at the top of the fresco; they are rolling back the picture plane as if it were so much wallpaper (detail, Figure 3.2). Here Giotto reminds us that all depictions of soul-space are ultimately illusions. Just as Dante

FIGURE 3.2. Detail from Arena Chapel *Last Judgment*. For Christian medievals, all representations of Heaven were ultimately an illusion. Here, an angel rolls back the image like so much wallpaper, revealing a glimpse of the "true" reality beyond—the pearly gates themselves.

understood that Heaven is beyond language, so Giotto knows it is beyond pictorial representation. Medieval depictions of soul-space, especially of Heaven, were never meant to be taken literally; they were always *metaphorical*. But if art could never capture the true reality of Heaven, it could at least point the viewer in the right direction. Thus, through the rents in the image we catch a tantalizing glimpse of the true reality beyond—two jeweled doors, the pearly gates themselves.

Precisely because Heaven was a wholly *other* plane of reality to the physical world, it could be dealt with in quite a separate fashion by late medieval artists. In depicting the Kingdom of God, Giotto felt under no obligation to adhere to the techniques of three-dimensional verisimilitude he had so carefully developed in the rest of the Arena Chapel. Today, it is the Christ cycle that most attracts attention from art historians, for here we see an early flickering of our own worldview, but to Giotto's contemporaries the more "old-fashioned" imagery of *The Last Judgment* was no less real. The stylistic dualism observed in the Arena Chapel reflects a worldview that took seriously the reality of both body *and* soul.

Crucially, this *metaphysical* dualism (so central to the medieval world picture) was mirrored in their cosmology, where it was expressed as a fundamental distinction between *terrestrial space* and *celestial* space. As we have already remarked, for medieval Christians the celestial realm was qualitatively different from the terrestrial realm. This distinction was central to their world picture, for while the earth was the realm of the mortal and mutable, things in the celestial realm were believed to be immortal, immutable, and eternal. Objects in the terrestrial realm were understood to be *transient*, like the human body; but those in the celestial realm were believed to be *permanent*, like the human soul. One of the great strengths of the medieval world picture was precisely this double parallel between the *metaphysical* dualism of body and soul, and the *cosmological* dualism of terrestrial and celestial space. Indeed, the latter dualism was seen as a reflection of the former.

Because medieval celestial space was qualitatively different from terrestrial space their cosmos was inherently *inhomogeneous.* In this respect their cosmology stood in stark contrast to modern scientific cosmology, for today the universe is seen to be essentially the same everywhere. This homogenous vision, as we shall see, is a direct extrapolation from the view of space developed by the Renaissance painters, for in the long run the power of "geometric figuring" would be extended to the stars. The purpose of this chapter is to trace this transition from the medieval to the modern vision of celestial space.

The fact that medieval celestial space was *not* the same as terrestrial space enabled them to see the celestial realm as a *metaphor* for the spiritual realm. It is no accident that the word "heavens" applied both to the domain of the stars and to the domain of God. This linguistic coincidence is in fact a common feature of many languages. "In Hebrew, Greek, and the Germanic and Romance languages, the same word denotes the divine heaven and the physical sky."[2] (English is quite rare in having a separate word for the physical sky.) As long as the medievals continued to see celestial space as qualitatively different, the heavenly bodies of the planets and stars could continue to serve as a pointer to the spiritual Heaven of God and the angels. Indeed this was the very basis of the great consonance between medieval cosmology and theology so beautifully articulated in Dante's *Paradiso*.

But what if celestial space was *not* different from terrestrial space? What if the two realms were not qualitatively different, but just parts of one continuous domain? What then would become of the glorious medieval holism? Just such a question was implicitly raised by the new conception of space pioneered by the perspective painters. Where the medieval cosmic system had been built on the belief that space is inherently inhomogeneous and hierarchical (as manifest in Dante's hierarchy of celestial spheres), the new perspectival space was quintessentially homogenous. In such a space there could be *no* inherent hierarchy because every place is the same as every other; no place is more special than any other because *all* are equal. The question thus arose: How far out from the earth might the perspective painters' Euclidian vision of space extend? Might this nonhierarchical space reach out to the stars themselves? Might terrestrial and celestial space constitute *a single* homogeneous realm? This question, at the heart of this chapter, was one of the major philosophical issues of the sixteenth and seventeenth centuries.

Today when we have put men on the moon and taken close-up pictures of Mars and Jupiter, the continuity of terrestrial and celestial space has become a "fact" of life. The engineering of the rockets that took the Apollo astronauts to the moon was predicated on the homogeneity of space, as will be any future missions to the planets and beyond. But if NASA engineers take the spatial continuity of the universe for granted, we must not forget that this also was a new idea in Western history. Even more so than the notion of the void itself, the idea of homogeneity between the earth and the stars seemed at first utterly incredible. The homogeneous universe might indeed be seen as one of the prime inventions of the modern scientific imagination, a concept so explosive it finally shattered the crystalline bubble of the medieval cosmos that had endured for a thousand years.

Once again the seeds of this cosmological revolution were presaged in the visual revolution of perspective painting. Again, it was the painters who paved the way for the scientists. This time our torchbearer is that great master of the early sixteenth century, Raphael. Consider Figure 3.3, the magnificent *Disputa*, painted for Pope Julius II in what are now the Raphael Rooms of the Vatican Palace. Here also, Raphael has rendered an image of the Christian Heaven, but unlike in Giotto's *Last Judgment*, he has sought to *unify* this divine space with earthly space. As we see, the image consists of two levels, the upper half representing Heaven, the lower half earth; between them is a robust bank of clouds. On the earthly level a phalanx of bishops, popes, and saints are arrayed in a semicircle on a marbled terrace; above them, seated on the matching semicircular cloud bank are Christ, the Virgin Mary, and John the Baptist, flanked by the apostles. Behind Christ's throne stands God, surrounded by angels. While the content of this image was wholly conventional, its form was anything but. As Edgerton explains, Raphael presented to "his papal patron an updated vision of the traditional [Christian] cosmos according to the latest conventions of linear perspective."[3] In other words, the artist had taken the extraordinary step of combining Heaven and earth in a single Euclidian space. Here, the spatial integrity of the perspectival image united the realms of God and of man.

If we ignore for a moment the uppermost portion of the

image—the part above Christ's throne where God and the angels commune—we see that from Christ down to the marble terrace the two realms are conjoined in a perspectively coherent image. Although the heavenly and earthly regions are delineated by the bank of clouds, both are depicted within the *same* Euclidian space. The integrity of the two regions is further signaled by the use of earthly naturalism in the heavenly realm, where, for example, heavenly feet cast shadows on the clouds. This is in stark contrast to the vision presented in *The Divine Comedy*, where Dante stressed that souls in the other world cast no shadows. Unlike Dante's Heaven, which was pointedly *not* natural, Raphael's heav-

FIGURE 3.3. Raphael, *Disputa (Dispute Concerning the Blessed Sacrament)*. Here Raphael attempted to unite Heaven and earth in a single homogeneous space.

enly domain seems just like another layer of terrestrial space; here the earthly "laws of nature" apparently still hold.

In the *Disputa,* both Heaven and earth are literally depicted from the same *point of view,* the perspectival "center of projection," which is positioned at the monstrance on the altar. In this Renaissance masterpiece, Edgerton tells us, Raphael "nearly succeeded in geometrizing medieval theology."[4] He nearly succeeded in bringing Heaven under Euclidian control. Almost, but not quite. For when we look at the uppermost portion of the image the spatial homogeneity suddenly breaks down. There with God in the Empyrean, we are back in the realm of Gothic symbolism, for here Raphael, like Giotto, abandons Euclidian space, and plunges us into a golden-rayed phantasm swirling with angelic spirits. Just as with Giotto, Raphael understood that in the true Heaven of the Empyrean, geometry must be jettisoned. In its ultimate, spiritual sense, the Christian Heaven *cannot* be unified with earthly physical space.

The medievals with their spiritually graded cosmos and their metaphysical dualism had innately understood that different levels of reality require different spatial domains: A *multiform reality demands a multiform conception of space.* Body and soul each need their own spatial milieu. But it was just this spatial dualism that was now being challenged by the perspective vision. With a homogeneous conception of space, how can there be *two* levels of reality? Homogeneous space by its very nature can only sustain *one* kind of reality. Thus the new artistic style that had originally been developed as a way of proselytizing on behalf of Christianity was now threatening the very basis of the Christian world picture.

Moreover, Raphael's *Disputa* was not intended to be just an image of the theological Heaven; Edgerton has discovered that the composition of this work encodes the precise structure of the *celestial* heavens as understood by astronomers of the time. Thus, the *Disputa* also suggests an implicit unification of terrestrial and celestial space. With the "logical art" of perspective Raphael was thereby calling into question one of the medieval church's "most cherished pronouncements about the composition of the cosmos."[5] Indeed, his attempt to reconcile earthly space and heavenly space raised "questions that would vex scientists for the next two hundred years."[6] Raphael was by no means the first person to wrestle with the spatial relationship between the heavens and the earth; rather the *Disputa* was an artistic encapsulation of one of the most consuming questions of the age: Just what is the nature of "heavenly" space, both in its *theological* and *celestial* senses? In trying to understand the latter, astronomers and scientists would gradually articulate a radical new cosmology.

The first person to mount a serious scholarly challenge to the medieval distinction between terrestrial and celestial space was a fifteenth-century contemporary of Hasdai Crescas. As with Crescas, Nicholas of Cusa was also a man whose ideas about space were far in advance of his time. Half a century before Raphael, Cusa addressed in his science the same questions the painter would attempt to resolve in his fresco, for like Raphael he too wanted to unify the heavens and the earth. A humanist, a philosopher, and a cardinal in the Roman Catholic Church, Cusa was in many ways the ideal Renaissance man. He collected ancient manuscripts, founded a hospital, and was a pioneer of experimental and theoretical science. His study of plant growth is considered "the first modern formal experiment in biology."[7] He was one also of the first champions of mathematically based science, and thus a precursor to the physicists of the seventeenth century. According to historian Eduard Dijksterhuis, his conclusions in this respect were so far-reaching that "a revolution in thought would have resulted had they been adopted and put into practice in the fifteenth century."[8]

Cusa's scientific speculations ranged across many fields, but it is for his cosmological ideas that he is most remembered today.

Although he lived a century before Copernicus, Alexander Koyre tells us that Cusa's cosmology went "far beyond anything that Copernicus ever dared think of."[9] Yet the starting point of his work was not any new astronomical data, but God. In this sense, he may properly be seen as "the last great philosopher of the dying Middle Ages." And like a star that ends with the explosion of a supernova, he was a spectacular finale to that glorious and too much maligned age.

Cusa laid out his cosmology in a curiously beautiful treatise entitled *On Learned Ignorance,* a book that on first encounter seems more like the ravings of some graceful alien than a work of "science." He began with the insistence that God alone is absolute, reading in this proposition a denial of all absolutes in the physical world. From this basis Cusa drew the conclusion that the universe has neither an outer boundary nor a center, since either would constitute an absolute. With this simple but extraordinary move Cusa demolished the medieval cosmos—for without an outer boundary the universe becomes necessarily an endless *unbounded* space. With one blow, then, the cardinal from Kues shattered the medieval "world-bubble" and released the cosmos from the crystalline prison of its celestial "spheres."[10]

Now, by definition, endless unbounded space cannot have a *center;* thus Cusa insisted that the earth was *not* the center of the cosmos. And neither was any other celestial body. In the endless space of the Cusan cosmos all positions were *equal.* Rejecting the medieval notion of a celestial hierarchy, Cusa asserted that "it is not true that the earth is the lowest and the lowliest" body in the universe.[11] On the contrary, it is a *star,* "a noble star which has a light and a heat and an influence" of its own.[12] In no uncertain terms, Cusa denied the medieval dualism of terrestrial and celestial space. His cosmos was a *unified realm* where nothing was lower or higher than anything else. As he put it: "There is one universal world."[13]

It is one of the more curious distortions of history that the displacement of humanity from the center of the cosmos is often said to have been a *demotion* for mankind. Yet nothing could be further from the truth. By shattering the heavenly spheres and breaking the medieval cosmic hierarchy, Cusa *elevated* the earth from the gutter of the cosmos and set it in the domain of celestial nobility. Not only Cusa but many latter cosmological innovators also saw the abandonment of the geocentric system as an enhancement of humanity's cosmic status. We should never forget that in the medieval system, the center was also the *bottom* of the cosmological scheme. Releasing the earth from this singular position could only mean a major cosmic promotion.

For Nicholas of Cusa, cosmic homogeneity was in fact a general principle. He boldly declared that wherever a man might be placed in the universe it would look the same: No place would present a special or unique view. Again this was in stark contrast to the medieval vision, where each celestial body, with its own appointed place, necessarily presented a unique perspective. With Cusa we thus have in crude form the first expression of an idea that has since become fundamental to modern science, the so-called "cosmological principle." According to this principle, the universe is essentially the same at *every point,* a requirement that underlies the modern belief in the repeatability of experiments. According to modern physics, it does not matter if I am on earth, or on Mars, or on Alpha Centauri, the same laws of nature will hold. Local conditions may vary, but cosmic homogeneity guarantees that the entire universe functions by the same natural laws. It was Cusa's assertion of this principle (albeit in rudimentary form), that, according to Jammer, gives us "justification for regarding Nicholas of Cusa as marking the turning point in the history of astronomy."[14]

There is a further aspect of Cusa's cosmology that also warrants our attention. The boundless space that he conceived he did not hesitate to fill with countless other stars. In contrast to the finite medieval cosmos, Cusa insisted that the universe has "worlds" without number. Moreover, he said, each of these other worlds is *inhabited:* "Natures of different nobility proceed from [God] and inhabit each region."[15] In fact, Cusa tells us that "none of the other regions of the stars are empty of inhabitants."[16] That is, the whole universe is populated. And just as in Cusa's cosmos there was no hierarchy among the celestial bodies, so also he asserted there was no hierarchy among their inhabitants. Whatever the nature of these other celestial beings, humans were not to be considered less noble than they.

Cusa's rejection of a hierarchy of celestial beings was nothing less than a refutation of the medieval hierarchy of angels. And here again we see that humanity was resoundingly elevated, raised up from cosmic guttersnipe to the rank of celestial being. Although Cusa admitted that the inhabitants of the sun and moon might be more "spiritlike" than the inhabitants of the earth, he categorically denied they were of a higher order.[17] Indeed, he went on to speculate that these celestial beings might also be subject to *death,* a phenomena hitherto confined to earthly creatures. "Death seems to be nothing except a composite thing's being resolved into its components. And who can know whether such such dissolution occurs only in regard to terrestrial inhabitants."[18]

In Cusa's cosmological scheme, it was not man, but the angels who were demoted; for they now became the potentially mortal equals of humans. Historically, this may be seen as the first step in a process that would culminate in the modern idea of *aliens.* What are ET and his ilk, after all, if not incarnated angels—beings from the stars made manifest in flesh? Like angels—good and bad—the aliens of modern science fiction are endowed with supernatural powers. Descending from the heavens in glowing orbs of light, they enter our lives pulsing with promise and visions of faraway paradise. The quintessential angel-aliens are the glowing humanoids of Steven Spielberg's *Close Encounters of the Third Kind*, beings who not only emanate heavenly light but communicate via music, a technological "harmony of the spheres." Or we might cite the radiant beings of Ron Howard's *Cocoon*, who promise the humans who go with them a life beyond sickness and suffering.

On the other hand, there are demonic aliens, viz. *Alien*, *Independence Day*, or any of the abduction scenarios that fuel American paranoia in the late 1990s. Classically, demons were fallen angels, and today our celestial brethren also come in the two flavors. Whether good or bad, aliens bear the burden of dreams once incorporated into the Christian world picture in the form of angels. By denying the celestial hierarchy on which these spirit beings depended, Cusa began the process of bringing these heavenly creatures *down to ground* and folding them into the web of *nature*.

Given the extraordinary scope of Cusa's conclusions, he can hardly fail to inspire our admiration, especially when we remind ourselves that he died a century and a half before the invention of the telescope. But as with Crescas, Cusa's work failed to influence most of his contemporaries. Only much later would his insights be recognized, and the innovative nature of his cosmology be fully appreciated. That Cusa could go so far without the aid of new instruments was a testament not only to his agile mind, but also to the changing character of the age itself.

Renaissance—rebirth—this was the word European scholars and artists used to describe the great cultural effluorescence of the time. "Man" was now being celebrated as the measure of all things, and as the sixteenth century dawned people were becoming increasingly dissatisfied with the lowly place in the cosmic hierarchy assigned to them by the medieval scheme. How, in an age that produced the magnificence of Michelangelo and Raphael, could people continue to believe that their rightful place was the *gutter* of the cosmos? Even without Cusa, Christian Europe was ripe for a change, and in this new century the tectonic plates of Western cosmology finally began to shift.

The prime mover of that shift was not Nicholas of Cusa, but an obscure Polish canon named Nicolas Copernicus. At the very time Raphael was painting the *Disputa* in Rome, Copernicus was a student in the north of Italy at the University of Bologna. Like Raphael, the Polish astronomer also would seek a unified "picture" of the heavens and the earth—indeed, he would devote his life to the task.

As with painting, the new science would be inspired by the burgeoning Renaissance spirit; for science, as always, is a cultural project. Chief among the "specific characteristics of the age" that fertilized the ground for Copernicus were the great sea voyages just beginning "to excite the imagination and avarice of Europeans."[19] Fifty years before his birth the Portuguese began to make voyages along the African coast, and just before the young Pole turned twenty, Columbus landed in America. As Thomas Kuhn has remarked, "successful voyages demanded improved maps and navigational techniques, and these depended in part on increased knowledge of the heavens."[20] Vast sums of money were at stake on these voyages, yet merchants and monarchs alike were at the mercy of their navigators, who in turn were at the mercy of the stars. In short, in order to plunder the gold and riches of the New World, the Old World needed a better understanding of astronomy.

If navigation was one inspiration for a new look at the stars, another was the urgent need for calendar reform. Because the date of Easter (the premier event in the Christian calendar) is determined by the cycles of both the sun and moon, astronomical accuracy was of considerable importance to the Roman Catholic Church.[21] In the sixteenth century, under Pope Gregory XIII, calendrical reform became an official church project and at one point Copernicus himself was asked to advise the papacy on the subject. The young Pole declined, however, on the grounds that the current understanding of celestial motions was so bad that no reform of the calendar could be undertaken until astronomy itself was reformed. That was the task that Copernicus made his life mission.

In the history of science there is no more boring "revolutionary" than Nicolaus Copernicus. After studying medicine and canon law at the universities of Bologna and Padua, this son of a patrician copper merchant from the town of Torun was appointed as a secular canon in the Cathedral of Frauenburg, a small city on the outer edge of Christendom in what is now part of Poland. Following a brief interlude as his uncle's physician, he settled into "this remote corner of Earth" and there he spent the rest of his days living what one commentator has called the leisurely "life of a provincial noblemen."[22] There were no great patrons, no glittering courts, no heated feuds; just a well-fed, self-satisfied country life.

Copernicus' duties helping to administer the cathedral's estates were not onerous—he and the other canons levied taxes, collected rents, and administered the local law. During his plentiful leisure hours he was free to devote himself to the study of the stars. There, in his private tower overlooking the lagoon of Frisches Haff, he spent thirty long years wrestling with the problem of celestial motion. The question that concerned him was this: How exactly do the sun, the moon, and the planets move through the sky? Here we witness one of the earliest of the new scientists tentatively reaching out with "virtual eyes" to probe the space around him. Reaching out, as it were, beyond his body to the remotest objects in the universe.

The problem facing Copernicus was that for all the philosophical beauty of the medieval cosmic system, the associated astronomy was not very accurate. In conjunction with the ancient world's cosmological system of heavenly spheres, the late medievals had also inherited the ancient *astronomical* system developed by Ptolemy of Alexandria in the second century. As the last major astronomer of the ancient world, Ptolemy had worked out a complex geometrical account of the motions of the celestial bodies, one that had been used by navigators ever since. This Ptolemaic system could be used to predict the positions of the sun and moon, the stars and planets, but it was far from accurate, and ships were constantly getting lost at sea, along with their precious cargoes.

In Ptolemy's description, the cosmic system was like a vast, clunky, celestial clockwork. It explained the movements of each celestial body in terms of a complex set of circular motions. While it is true as a first approximation that the celestial bodies travel in circles around the earth, on closer examination their paths are *not* perfect circles. Some of the planets have especially warped paths. In order to explain these deviations from circular perfection, ancient astronomers had hit upon the idea that each celestial orbit must be the result of several circular motions combined together. One can imagine this as a set of celestial gears with each major gear supplemented by additional smaller gears. Just as a wind-up ballerina can be made to perform a dance by a complex arrangement of gears, so the ancients reasoned that the celestial dance of the stars and planets could be explained by complex arrangements of circular motions. Ptolemy's system was the pinnacle of this process.

But according to Copernicus, Ptolemy's system was ugly. The young canon could not believe God would have created such an aesthetically awful system. Thus while Copernicus was certainly motivated by practical considerations vis-à-vis navigation and calendrical reform, he was also inspired by *aesthetic* concerns. In particular, says historian Fernand Hallyn, he was inspired by the aesthetics of the Renaissance painters, by their ideals of beauty, harmony, and symmetry.[23] "The principle consideration" of astronomy, Copernicus wrote, is to deduce "the structure of the universe and the true symmetry of its parts."[24] Indeed, the vision of the cosmos he developed might well be seen as the ultimate Renaissance picture of the world.

In searching for a more "harmonious" and "symmetrical" vision of the cosmic system, Copernicus came up with the idea of a *sun-centered cosmos*. Here, the sun replaced the earth as the focus of the system. It is beyond the scope of this book to describe how Copernicus came to this extraordinary conclusion, but suffice it to say that he was not the first.[25] A heliocentric cosmic system was considered by a number of people in the early sixteenth century, and the idea had in fact been known to the ancient Greeks two thousand years before. What Copernicus *did* do was to work through the laborious details of how a heliocentric system might really work. In contrast to Cusa, who never articulated the details of his cosmic system, Copernicus slogged through the geometry to show just how the planets might actually move in a sun-centered system. For this monumentally tedious task, modern cosmology owes him an enormous debt.

Much has been made in popular science books about the supposed simplicity of the Copernican system, but nothing could be further from the truth. Harvard historian Owen Gingerich has shown that the Copernican cosmos was neither simpler nor more accurate than its Ptolemaic predecessor.[26] On the contrary, it was just as complex and just as inaccurate. Copernicus too described the celestial motions with a Byzantine collection of invisible celestial gears—his account of the orbit of the earth, for example, required no less than nine celestial circles. In this respect, his system was just as ugly as its predecessor. What could not be overlooked was the fact that Copernicus' system was *no worse* than its predecessor. Thus, for the first time in Western history, the geocentric vision of the universe had a serious competitor. From now on, heliocentrism would have to be accepted as at least a potential option.

Yet if Copernicus' system was no simpler than Ptolemy's, with respect to the history of space it did have several important advantages. Firstly, in a heliocentric system the earth became one of the planets, and so, as with Cusa's cosmology, man was again catapulted into the realm of *celestial space*. The second advantage of Copernicus' system was that by setting the earth in motion around the sun, the stars could become stationary. As we have seen, in the geocentric system the stars revolved around the earth on a vast crystal sphere. From a theological perspective that might have been acceptable, but practically speaking it seemed a little absurd. Copernicus himself noted that it made much more sense for a relatively small body like the earth to be moving than for the vast sphere of the stars.

Indeed for all Copernicus' reputation as a modern, he still believed in heavenly celestial spheres. His cosmology actually *demanded* them, for in his system God remained the source of celestial motion. Vis-à-vis the history of space, there is also the issue that while Copernicus set the earth among the planets, he by no means destroyed the distinction between terrestrial and celestial space. To do so, he would have had to believe the celestial bodies were composed of solid physical matter like the earth. But unlike Nicholas of Cusa, who was beginning to think along these lines, there is no evidence Copernicus believed any such thing. In the Polish canon's cosmological vision, the celestial realm remained an *ethereal* "other" realm, and in so many ways he was more a medieval than a modern thinker. Plainly put, says Kuhn, "the Copernican revolution as we know it, is scarcely to be found" in Copernicus himself.[27] Yet whatever Copernicus' personal beliefs, there is no doubt that his work ushered in a new era in cosmological thinking.

The man who truly saw the potential of the heliocentric spatial scheme, and who truly demolished the medieval distinction between celestial and terrestrial space was not Copernicus, but the German mathematician Johannes Kepler. The "great men" of any field always inspire biographies, but few in the history of science inspire love. One of those is Kepler. A sickly runt from the German town of Weil-der-Stadt, born into a family of dissolutes and drunks, Kepler rose from these squalid beginnings to become one of the premier scientific geniuses of any age. When Newton said, "If I have seen further it is by standing on the shoulders of giants," he referred to no one so much as Kepler. Kepler's laws of planetary motion would pave the way for Newton's discovery of the law of gravity and, with that, the final union of celestial and terrestrial space.

Nothing in Kepler's pathetic childhood seems to have prepared him for such a momentous role. Aside from the trauma of his brawling family, he was so unpopular among his classmates that "his fellows regarded him as an intolerable egghead and beat him up at every opportunity."[28] Sensitive, sickly, and overtly religious, young Kepler made an easy and inviting target. Reminiscing about his tortured childhood, he once wrote: "As a boy of ten years when he first read Holy Scripture . . . he grieved that on account of the impurity of his life, the honor to be a prophet was denied him."[29] But Kepler *would* become a "prophet," for he was the first true *astrophysicist*—the first person to see the celestial realm as a space of concrete physical action.

Kepler took Copernicus' sun-centered vision, but he threw out the old canon's medieval methods and set himself the task of explaining how a heliocentric system might really physically work. In this respect he took the decisive step that Copernicus never dared make: He regarded the celestial realm as a concrete *physical domain* (just like the terrestrial realm), and he treated the celestial bodies as concrete *material bodies* that must function according to natural physical laws. With this intuition, the weiner from Weil-der-Stadt reinvented the world.

Once again, to we who have witnessed men walking on the moon and robot probes crawling over the surface of Mars, Kepler's intuition may seem rather mundane. But it cannot be overstated what a giant intellectual leap this was. When Neil Armstrong walked on the moon he was, in effect, following in the footsteps of Johannes Kepler. We humans could not even *dream* of treading on the lunar surface until we had come to see the moon as a concrete physical place, and Kepler was the first person to do so. In fact, three centuries before the Apollo missions, he actually envisaged a physical voyage to the moon.

Because Kepler believed the celestial realm was a concrete physical domain, he was at last able to free his mind from the old Ptolemaic methods to look for a genuine alternative to the hackneyed system of celestial gears. By doing so he discovered that the planets move around the sun not in some complex combination of circles—as every Western astronomer since Aristotle had insisted—but in *ellipses*. He found that the path of each planet is in fact an elegant ellipse with the sun at one focus. In this sacrilegious deformation from circular perfection lay the foundations for a genuinely postmedieval cosmology. According to Kepler, it was not God that propelled the planets around their orbits, but *physical forces* inherent in the cosmic system. For him, the problem of celestial motion was not a matter for theology, but for *physics*. For this reason he can truly be regarded as founder of that quintessentially modern science, "astro-physics." Specifically, Kepler said, the planets are moved around their orbits by a physical force that emanates from *the sun*.

In the history of Western cosmology, this ranks as one of the premier insights. That Kepler's name is not known as far and wide as Copernicus' is one of the greater injustices of popular history. What we have here with Kepler's solar force and his planetary ellipses is the folding of the celestial domain into the realm of *natural science*. As the first person to propose the existence of natural physical forces and laws operating in the celestial realm, Kepler issued the definitive challenge to the medieval distinction between celestial and terrestrial space. His universe was not only unified, it was physically viable throughout. In this sense, it is he, not Copernicus, who is the first of the true "moderns."

Kepler's commitment to the unity of celestial and terrestrial space is evident in a curious little book now regarded as the first work of science fiction. Entitled simply *Somnium (The Dream)*, Kepler here describes an imagined trip to the moon. Eschewing all hints of medievalism, his moon is a solid material orb, like the earth. On it there are mountains and caves, oceans and rivers; plants grow, animals are born, and they die. With this moon we are definitively in the realm of *nature*. According to Kepler's story, our lunar cousin is populated by a motley collection of lizard-like creatures. "In general," we are told, "the serpentine nature is predominant." As the narrator describes these creatures, "they roam in crowds over their whole sphere, each according to his own nature: some use their legs, which far surpass those of our camels; some resort to wings; and some follow the . . . water in boats."[30] In other words, some of them are intelligent. They even have a basic grasp of astronomy. Here in the early seventeenth century, we see then the logical culmination of the movement begun by Nicholas of Cusa two hundred years before. The grounding of the angels is now complete; encased in flesh, the beings of the stars have become *mortal* creatures. Here is the modern "alien" in full-bodied, solid material form.

What a radical transformation has taken place since Dante ascended into the heavens three centuries earlier. Gone now are the crystal spheres and angelic harmonies of *The Divine Comedy*, replaced by scuttling serpents who hide in caves and shed their skins like husks. Gone now is the "singing silence" of the medieval Heaven, to be overtaken by "scientific progress" and by visions of intelligent saurians. Exciting though it may be to witness the birth of science fiction, one cannot help but feel a hint of sadness at this new physicalist vision. From here on, celestial space will ring not with the songs of cherubim and seraphim, but with the roar of rockets and the woosh of warp drives.

With Kepler's *Somnium*, Western culture reached a critical junction, for there is no question that lunar lizards sounded the death knell for the medieval world picture. To put it at its most basic, celestial space cannot sustain angels *and* boat-building serpents. You cannot have it both ways: Either the celestial realm is a metaphor for the spiritual space of Heaven, a space populated by "angels," or it is a physical space filled with material planets inhabited by "aliens." Although no one would be formally asked to choose, we all know which way the vote would go. As a pointer to the future, the lizards stand as a peculiarly apt precursor of things to come.

Kepler, more than anyone, formulated the modern vision of celestial space as a concrete physical realm, but while he crystalized this vision it was Galileo who propelled the idea to the forefront of Western consciousness. Thus it is *his* name, not Kepler's, that is usually associated with this seminal step. The key to Galileo's success was an astounding new instrument. While Kepler was traveling to the moon in his mind, Galileo was busy peering at it through a telescope, and like Kepler what he found there was not a misty "ethereal" orb, but mountains! Announcing his telescopic discoveries to the world, Galileo declared "the surface of the Moon to be not smooth, even, and perfectly spherical, as the great crowd of philosophers have believed . . . but on the contrary . . . it is like the face of the earth itself, which is marked here and there with chains of mountains and depths of valleys."[31]

In addition to mountains on the moon, the new "optick

tube" provided concrete evidence against the medieval belief in the immutability of the celestial domain. Through his telescope Galileo saw that the sun had spots which moved across its face. Thus change could occur in the heavens, as on earth. The mutability of the heavens was also suggested by the discovery that comets were not atmospheric phenomena, as Aristotle had argued, but genuine celestial residents. All in all, the evidence that came flooding down the optick tube increasingly pronounced in favor of the celestial realm as a concrete physical domain.

Ever since Galileo first pointed one at the moon, the telescope has become humanity's pipeline to the stars, the instrument through which we have been able to send our "eyes" roving out into celestial space far beyond what we can naturally see. If, as we saw in the previous chapter, perspectival imagery trained Western minds to see with a "virtual eye," the telescope extended our virtual gaze beyond the wildest imaginings of the Renaissance painters. Precisely because celestial space is not a place we can physically go (even the few elite astronauts have never been further than the moon), it is a space that in general we know only through "virtual eyes." In this respect, our experience of "outer space" parallels our experience of cyberspace, for *it* too is a space we do not experience physically. Both outer space and cyberspace are *mediated* spaces that we see through a technological filter. And just as today we are beginning to get a sense of the potential vastness of cyberspace, so also Europeans of the seventeenth century were just beginning to get a sense of the potential vastness of the new space they were discovering at the other end of their optick tubes.

By the middle of the seventeenth century, the European scientific community had more or less accepted that the cosmos was both sun-centered and physical. But one question remained a complete mystery: Just how big was our universe? The medieval cosmos had been small and finite, with a definitive boundary at the outermost sphere. Did the new heliocentric cosmos also have an outer boundary? Or might it go on forever? Might celestial space, in fact, be *infinite*—as Cusa had suggested two centuries before?

Surprisingly, perhaps it was this idea of infinite space that most caused upset. From a Christian theological perspective the notion of an infinite universe was particularly unacceptable because it implied a world *without form*. The whole Christian-Aristotelian synthesis had been grounded on the belief that in the architecture of the cosmos we can discern the reflection of a divine Creator. But how could God be reflected in formlessness? Kepler in particular argued against it. Of all the propositions of the new cosmology, this was the one that met with most resistance. The moving earth, the sun's central place among the planets, the materiality of celestial bodies—all these gradually became part of the scientific world picture in the decades after the invention of the telescope. What most people could not accept was an infinite formless void. Christian theology and Greek philosophy both rebelled against the infinite—the dreaded *apeiron*—and as with void space itself, acceptance of the idea of infinite space would require a significant shift in the Western mind-set.

After Cusa, the next major champion of infinite space was the heretical Italian mystic Giordano Bruno, who was burned at the stake in the year 1600. Following in the footsteps of Cusa, Bruno insisted that the universe was infinite and filled with countless other stars. No one, he wrote "could ever find a half-probable argument . . . that this corporeal universe can be bounded, and consequently that the stars which are contained in its space are likewise finite in number."[32] Paradoxically, given what we have just said about theological objections to the infinite, Bruno justified his unending universe by recourse to God. Here, he drew upon the theological tradition that envisioned the Christian deity as a god of abundance. In this tradition, "a larger and more populous universe must connote a more perfect deity"—or, to put it another way, an infinite God could "be satisfied only by an infinite act of creation."[33]

In promoting the idea of an infinite creation, Bruno specifically stressed that *space* itself was infinite. "We who see an aerial, ethereal, spiritual, liquid body . . . we know for sure that this [entity] which has been caused and initiated by an infinite cause and principle must be infinitely infinite."[34] Thus, in Bruno's cosmology infinite space became a direct reflection of an infinite God. It is just this *theologizing of space* that would finally make the *apeiron* palatable. During the late sixteenth and seventeenth centuries an impressive list of thinkers gradually constructed a theology of infinite space and justified this hitherto abominable concept by associating it with God.

One of the key people whose work contributed to the eventual acceptance of infinite space was René Descartes. Although Descartes himself rejected void space per se (preferring with Aristotle to see the universe as a plenum in which matter filled the entire volume), like Bruno's universe his was also infinite.

Despite his reputation as a hard-headed rationalist, Descartes' approach to science was founded on a mystical revelation that he believed had come to him directly from God. On November 10, 1619, while resting one night in an inn, the young philosopher had a vision, later followed by several dreams, in which he was visited by a higher power. In this vision, says Edwin Burtt, "the Angel of Truth appeared to him, and seemed to justify through supernatural insight, the conviction which had already been deepening in his mind, that mathematics was the sole key needed to unlock the secrets of nature."[35] From this angelic message, Descartes went on to envision his mechanistic world picture in which the universe consisted of matter moving through infinite space according to strict mathematical laws.

Descartes had carefully crafted this mechanistic vision hoping to support his Roman Catholic faith, but much to his despair, many people interpreted the Cartesian cosmos as a dangerous atheistical construct. The only role that seemed left for God in this universal machine was to supply the mathematical laws by which the system runs. To many of his peers Descartes seemed to have written God out of the universe in any meaningful way. How could a believing Christian accept such a "soul-less" vision of the world? Yet many scientists of the seventeenth century wanted to accept some form of mechanism. Like Descartes' many of them believed that the universe was, in some sense, akin to a machine. What they wanted, in effect, was a more Christian machine. In their quest for a more "spiritualized" version of mechanism, space would play a critical role.

One of the earliest attempts to spiritualize the Cartesian cosmos was carried out by the English divine Henry More, who set out to supplement Descartes' world machine with what he saw as specifically Christian features. More rejected Descartes' belief in a plenum, and following the ancient atomists he declared that the universe was made up of atoms and void space. Like Bruno, he justified this empty space by divinizing it, calling it a "subtile" substance, a "Divine Amplitude." For More, in fact, space was the mediating substance between physical matter and divine spirit, the link between the material and the spiritual realms. By such theological moves, More and his contemporaries sought religious credibility for a mechanistic universe. What they wanted was nothing less than a new fusion between science and religion, a mechanistic world picture that could be reconciled with their Christian faith. In this respect, as with so many others, the trajectory of the "scientific revolution" would reach its apogee with More's young colleague, Isaac Newton—the man who would sear the new cosmology, and the new conception of space, into the collective Western consciousness.

Isaac Newton: scientific genius, Christian heretic, practitioner of alchemy: All are accurate descriptions of the man who towers over the modern West's world picture in a manner that can only be paralleled with Aristotle's dominance of the ancient world. But to understand Newton we must look at more than his science, for above all he was motivated by religious inspiration. The depth of Newton's faith may be gauged by the fact that he was quite prepared to forego an academic career rather than swear fealty to a theological view he did not support. At the time, Cambridge University still demanded that academics be ministers in the Anglican Church, and ordination in turn required a declaration of belief in the Trinity—that core Christian doctrine which asserts that divinity takes three simultaneous forms: the Father, Son, and Holy Ghost. But Newton secretly adhered to the heresy of Arianism, which repudiates the Trinity and insists on the indissoluble unity of the Christian deity. He was not fool enough to make this heretical stance public, but neither was he willing to pretend allegiance to something he didn't believe. As the day of ordination drew near he was mentally preparing to leave Cambridge, when at the eleventh hour came a dispensation from the king: Newton could remain at the university without being ordained. The fact that he was a heretic would remain his personal secret.

This story is of a personal interest, for it casts an unusual light on a man so universally known for his science, but it also gives us a vivid glimpse of the contingency of history. What if the dispensation had *not* come, and Newton had left Cambridge for the life of a country squire—the destiny his parents intended for him? Would he have still pursued his science? Would there have been a *Principia*, a unifying "bible" to tie together the new cosmology? Newton's biographer Richard Westfall has considered these questions and with respect to the latter has concluded that in all likelihood there would not.[36] This preternaturally inquisitive mind would no doubt have continued to think about the world around itself, but without the setting of Cambridge his ideas would probably not have been published. And without the thunderous impact of the *Principia*, the Western history of space might well have been quite different.

What Newton presented to the world in his legendary (and legendarily difficult) tome was an overarching synthesis that tied together the cosmological insights of all his major forebears, notably Copernicus, Kepler, Galileo, and Descartes. With giants like these on whose shoulders to stand, Newton was positioned for a far-reaching view of celestial space and he does not disappoint us.

First and foremost, Newton completed the unification of celestial and terrestrial space that Kepler had begun. The key to his synthesis was a simple mathematical equation that even today stands as the archetypal "law of nature"—the "the law of gravity." Originally inspired, as legend would have it, by the fall of an apple from a tree in his mother's garden, Newton went on to show that the same force which caused the fruit to fall to the earth could also explain how the moon revolved around the earth and how the planets revolved around the sun. Indeed, the same force that Kepler had speculated as holding the planets in their orbits, Newton now demonstrated was also responsible for keeping our feet anchored to the ground. A single physical force thereby operated in both the celestial and terrestrial realms.

Moreover, hidden in the law of gravity was a metaphysical bombshell about the nature of the celestial bodies. The essence of Newton's law is a force of attraction between two *physical masses*. Where there is gravity there must be *matter*, raw solid physical matter. Now as Kepler's laws of planetary motion testified, Newton's gravity operates in the *celestial domain*—the elliptical shape of the planetary orbits is a direct consequence of Newton's law. With a gravitational force operating between the sun and the planets, these celestial bodies *must* therefore all be concrete material bodies like the earth!

It is a little-recognized aspect of the scientific revolution that the physicalization of the celestial realm was ultimately clinched by the ephemerality of a mathematical equation. Before Newton's equation, people could continue to argue about the constitution of the celestial bodies, but after the law of gravity had been discovered, that battle was effectively over. Matter now reigned supreme, not just on earth but throughout the cosmos. With his law, Newton thus completed the revolution that Cusa had first imagined: Celestial space and terrestrial space were now united as one *continuous physical domain.*

But unlike Descartes' universe, Newton's was imbued with divine spirit, for, following Henry More, he too associated space with God. Indeed, for Newton, the very presence of God was synonymous with the presence of space. As he wrote, God "endures for ever, and is everywhere present; and by existing always and everywhere, He constitutes duration and space."[37] More so even than his predecessors, Newton justified his vision of space on theological grounds. Space, as he famously put it, was God's "sensorium"—the medium through which the deity exercised His all-seeing eye and His all-encompassing power. For Newton, the presence of God within the universe was indeed *guaranteed* by the presence of space. And because in his view God was *everywhere,* then space must also be everywhere—and hence *infinite.*

Over the course of two centuries the unthinkable had thus become acceptable: An infinite formless universe pervaded by infinite void space had become the basis of Western cosmology. First people had come to accept the idea of void space itself, then they had accepted the celestial domain as a concrete physical realm, and finally they had come to accept that this realm extended to infinity. And all this they had justified on religious grounds.

In the long run, however, while the divinization of space had been psychologically necessary to overcome initial resistance to infinity (and to the void itself), a theological view of space was not in truth necessary to the new cosmology. Thus in the eighteenth century, after Newton's death, we witness the spectacle of less religiously inclined scientists stripping away the theological frills from his system. By the middle of that century the new cosmology had been almost totally secularized, and it is this essentially atheistic Newtonianism that has come to dominate the modern West. In the end, the anti-Cartesians were right: Mechanism leads almost inevitably to an atheistic world picture. Despite the efforts of More and Newton, space proved an insufficient medium for the perpetuation of a deity within the cosmic system. In the final analysis, the materialists won the day, and in the Age of Reason man stood not at the center of an angel-filled cosmos with everything connected to God, but on a large lump of rock revolving purposelessly in an infinite Euclidian void. The medieval era was now truly over.

Let us stop for a moment to reflect on the momentous changes that have been described in this chapter. Popular histories of science would have us believe that with the new cosmology humanity had "progressed" from ignorant darkness to the glorious light of "truth." The "true" architecture of the cosmos had supposedly been discovered, as humans finally "knew" where they stood in the cosmic scheme. Just as the sun had displaced the earth at the center of the planetary system, so science displaced theology at the center of our intellectual system. With man's mind now revolving around this "true" source of light, the future was supposedly assured in an endless ascent toward Truth.

But while we shall see in the following chapter that modern cosmology has been extraordinarily successful, by going down this profoundly physicalist path Western humanity has also lost something of immeasurable importance. The very homogenization of space that is at the heart of modern cosmology's success is also responsible for the banishment from our world picture of any kind of spiritual space. In a *homogeneous* space only *one* kind of reality can be accommodated, and in the scientific world picture that is the *physical reality of matter.* In medieval cosmology, the accommodation of body and soul had been premised on the belief that space was *inhomogeneous.* By rendering obsolete the old division between terrestrial and celestial space, modern cosmologists forced their own metaphysical hand and reduced reality to just one half of the classical body-soul dimorphism. Moreover, once this physical space was itself extended to infinity, there was no "room" left for any kind of spiritual space.

To put this in its starkest terms, in the infinite Euclidian void of Newtonian cosmology there was literally *no place* for anything like a "soul" or "spirit" to be. In the medieval cosmos the soul's "place" was "beyond" the stars, for as we noted at the start of this work, with a finite universe it was possible to imagine — even if, strictly speaking, only in a metaphorical sense — that there was plenty of "room" left outside the physical world. But once the physical world became infinite, where could any kind of spiritual realm possibly be? By *unbounding* the physical realm, the Christian spiritual realm was thereby squeezed out of the cosmic system. That excision precipitated in the Western world a psychological crisis whose effects we are still wrestling with today.

It is important to note here that this is a specifically Western problem. The reason we *lost* our spiritual space, as it were, is because we had linked it to celestial space. We had "located" it, metaphorically speaking, up there beyond the stars. When celestial space became infinite, our spiritual space was thereby annihilated. Yet as Christine Wertheim has pointed out, if we had not located our spirit realm up there in the first place, this crisis would not have resulted. Many other cultures do not in fact tie their spiritual space to the starry heavens. Many so-called "primitive" people locate their spirit realm in dreams, or in a mythical past that remains interlinked with the present. For these cultures, the infinitization of celestial space would not necessarily have precipitated the crisis that it has done in the West. *They* could have had an infinite celestial space and still have *kept* their spiritual domain.

In the purely physical cosmology of Newton and his intellectual descendents there could of course be no place for the Christian Heaven and Hell. For the medievals, Heaven and Hell (though technically outside the universe), were woven into a scheme in which all of space was spiritually graded. In Euclidian space, however, one end of the universe is the same as the other, and Heaven and Hell become empty symbols. With no links to physical reality, these spiritual places were inexorably doomed to extinction. Indeed, from the late seventeenth century on, the new physicalist vision has been invoked as a powerful epistemic scythe to hack off *anything* that could not be accommodated into the materialist conception of reality. Increasingly over the past three centuries, reality has come to be seen as the *physical* world alone. Thus as I stated at the start of this work, it is a complete misnomer to call the modern scientific world picture dualistic; it is *monistic*, admitting the reality *only* of physical phenomena. Here, the Christian soul is not the basis for another level of reality, as the medievals believed, but a chimera of the imagination—Gilbert Ryle's "ghost in the machine."

Newton and Descartes would have been appalled at this desacralization of the scientific world picture, but this is the end result of the cosmology they bequeathed us. Whatever their personal beliefs, Newton's mathematical science and Descartes' dualistic metaphysics have ultimately served as stepping stones to a rampant materialist monism. Descartes in particular, with his radical divide between a physically extended realm of matter in motion (the *res extensa*) and an invisible realm of thoughts, feelings, and spiritual experience (the *res cognitans*) powerfully tilted the scales toward monism. Since the new science would describe only the *res extensa*, it was only this realm that would receive the sanction of scientific authority. As that authority increased, anything outside science's purview came increasingly under attack.

While Descartes himself insisted on the reality of the *res cognitans*, his radical exclusion of this immaterial realm from the methods and practices of science left it highly vulnerable to claims of "unreality." In the medieval world picture, the spiritual realm (what I have been calling soul-space) had been secured by its intimate entwining with the science and cosmology of the time. But with Descartes' dualism there were no links between the realm of matter and the realm of spirit. Without links to the concrete world of physical science, the Cartesian *res cogitans* quickly became (like the Christian Heaven) an empty symbol. Not surprisingly, it wasn't before long people were casting doubts on its entire existence.

The trend was set by the English philosopher Thomas Hobbes, who even in Descartes' lifetime declared that mental phenomena were merely secondary by-products of the primary reality which was matter in motion. "Mind will be nothing but the motions of certain parts of an organic body," he wrote, in what soon became a call to arms to the growing legions of materialists.[38] Thus like Heaven and Hell, the Cartesian *res cogitans* was also quickly annihilated from the realm of the real. By the end of the eighteenth century monism was in full swing. To quote the incisive words of Edwin Burtt:

> The natural world was [now] portrayed as a vast, self-contained mathematical machine, consisting of motions of matter in space and time, and man with his purposes, feelings, and secondary qualities was shoved apart as an unim-

portant spectator and semi-real effect of the great mathematical drama outside.[39]

For the first time in history, humanity had produced a purely physicalist world picture, one in which mind/spirit/soul had no place at all.

CHAPTER FOUR

RELATIVISTIC SPACE

In the Judeo-Christian book of Genesis it is God who creates "the heavens and the earth." In six seminal days, he molds the nascent "darkness" into the splendor that is Creation. Calling forth light, he separates day from night, then parting the "waters" above and below the "firmament" he delineates the celestial and terrestrial realms. At this point "the heavens" are empty, and it is not until the fourth day that the divine architect turns His attention to filling the cosmic void. Here is the Genesis account of celestial ontogeny:

> And God said, "Let there be lights in the firmament of the heavens to separate the day from the night; and let them be for signs and for seasons and for days and years." . . . And God made the two great lights, the greater light to rule the day, and the lesser light to rule the night; and he made the stars also.[1]

With the sun, the moon, and stars now setting the cosmic clockwork of seasons and years, the Christian Creator now applies His energy to bringing forth living creatures: first the fishes of the sea and the birds of the air, then the "cattle and creeping things and beasts of the earth," finally man and woman. Thus does

Genesis unfold its story of Creation, articulating in poetic language the bringing forth of a universe out of nothing. Everything has a beginning, and for the writers of the Old Testament the ontogenic force was God.

Unlike the Bible, however, Newtonian cosmology told no Creation story. While the laws of Newtonian physics could describe how planets revolved around their suns, and moons around their earths, these laws had nothing whatsoever to say about cosmological *history*.

The Newtonian cosmos did not *become*, it simply *was*. Moreover, while Newton's laws were silent on the subject of Creation, so also was empirical observation. When looking through their telescopes, astronomers of the eighteenth century could detect no sense of a cosmic story, no sense of a beginning, or indeed of an end. The view that came through the optick tube was of an apparently *timeless* universe. No matter how far men looked out in space, there was no sign that anything had ever been different. If God had created this cosmos—as Newton himself never doubted—then He seemed to have done His utmost to erase all trace of the generative process.

The fact that the new science did *not* have its own Creation story made it all the easier to harmonize with Christianity. In effect, there was no competition. One could readily agree with Newton that the Christian deity had made the world, and that when He did He had built into it the laws of physics. In other words, along with the heavens and the earth, one could believe that the six days of Genesis had also produced the laws of motion and gravity. For believing Christians of the early eighteenth century, the lack of a cosmic history in the scientific account of the universe was indeed a source of satisfaction: One could accept both the Bible and Mr. Newton without tension.

But as time wore on, scientists of a more philosophical bent began to find the ahistorical nature of the scientific world picture increasingly unsatisfactory. If the new science was to be truly successful, they believed, it would have to explain the question of origins; it would have to describe in its *own* terms how a universe could arise out of nothing. In this chapter we consider the question of the origin of the universe, and along with that the origin of physical space itself. As we shall see, the scientific answer to these questions would emerge from an extraordinary new conception of space developed in our own century—one that would eventually replace Euclidian space as the foundation of modern cosmology.

Beginning in the second half of the eighteenth century a number of people began to propose scientific theories of cosmic genesis. The most original and comprehensive of these visions came from the great German metaphysician and philosopher Immanuel Kant. Although he was a devout theist, Kant believed that planets, stars, and even whole star systems must arise from purely natural processes. In 1775, in his *Universal Natural History and Theory of the Heavens*, Kant described a process by which he believed that whole solar systems could condense out of clouds of cosmic dust. He imagined a huge disc of dust slowly rotating in space, which then condensed to form the luminous matter of the sun and the dark masses of the planets. Far in advance of his time, Kant even ventured to propose how entire galaxies might form, congealing out of enormous galactic-scale clouds. Later in the century these ideas were taken up by the supreme astronomer of the age, Pierre-Simon Laplace, who along with Kant believed that science could conjure a universe out of nothing more than raw matter and Newton's laws of motion.

But without any knowledge of stellar processes, Kant and Laplace's ideas were just speculative flights of fancy built on little more than their faith in the scientific method. "Armchair theorists," science writer Timothy Ferris has dubbed such men.[2] With no empirical grounding, this pioneering work on cosmic evolution was quickly forgotten, and for most astronomers of the nineteenth century the question of cosmic origin was one to be vigorously avoided. What was the point in speculating about a subject on which science could offer no empirical data? Much better to stick to questions that their instruments *could* illuminate; and with telescopic technology evolving by leaps and bounds there was no lack of interesting projects to pursue. Aside from the glory of the planets, a man with a good telescope might turn his attention to comets, or to the sun with its enigmatic spots and plumes. He might catalog the stars, which on close inspection turned out to come in a surprising variety of types; or an ambitious astronomer might study the nebulae, those fuzzy blobs of light which lurk at the edges of our Milky Way.

With so many fascinating phenomena to explore, most nineteenth-century astronomers were content to leave the issue of Creation alone. Those who chose to, could continue to imagine that God had somehow created the cosmos through as yet undefined processes, but by the latter half of the century many scientists preferred to simply take Newton's laws at face value and imagine that the universe had existed in much the same state since time immemorial. In place of the Christian Creation story, there thus emerged a de facto scientific picture of *cosmic stasis:* a universe without beginning or end, a cosmos that simply is. According to this picture, the universe was *without history,* an eternal timeless pattern of stars that had endured, and would endure, forever. Over the course of the nineteenth century this static picture became so deeply entrenched in most scientists' minds that by the early twentieth century the idea of a cosmic origin had become almost unthinkable in scientific circles.

In the 1920s, however, this static conception of the cosmos was shattered by a dashing young Missourian named Edwin Hubble, who discovered that distant stars are rushing away from us at immense speeds. It was as if they were compelled outward by some unimaginable force, flinging them through space like so much cosmic shrapnel. The implications of Hubble's discovery would change forever our conception of the universe. The old static Newtonian picture, with its rigid Euclidian space, would be replaced by a much more dynamic vision of the cosmos, and by a new *dynamic* conception of space, itself.

From an early age the man at the center of this cosmic upheaval seemed destined for great things. At least, that is what most of those around him thought, and Hubble himself seems to have little doubted his own abilities. Tall, good-looking, a gifted athlete and a superb scholar, Hubble had every reason to approach the future with confidence. Once started on his chosen career as an astronomer he would pursue an unerring trajectory. He seemingly possessed a built-in compass for asking the right questions at the right time. By 1921, when requested to provide a personal update for the community of Rhodes scholars, to which he belonged, he could happily respond:

> My one distinction is that of being the only Astronomer amongst the Brotherhood. Whilst toying with such minor matters as the structure of the universe, however, I sometimes tackle the serious problems of life, liberty, etc. and strive earnestly to circumvent these damnable prohibition laws.[3]

If Hubble's rather pompous reply grated with what one biographer has called an "affected tone," there is no doubt that "the structure of the universe" *was* occupying his full attention.[4] And here Hubble was to make not one, but several major contributions.

In the 1920s the idea of a static universe still dominated astronomical thinking, and Hubble, who in many ways was a rather orthodox thinker, never dreamed of overturning this cherished tenet of Newtonian cosmology. What led him in this heretical direction was no incipient radicalism, but the crisp data of his beloved nebulae. Taking their name from the Latin word for "fuzzy," nebulae are hazy patches of light that astronomers find scattered throughout the night sky. Most are only visible with telescopes, and since the eighteenth century scientists had debated the nature of these mysterious blobs. Some contained stars, but many seemed not to. What were these mysterious cosmological objects?

For most astronomers the answer seemed clear: They are luminous clouds of gas floating within the Milky Way. But along with this majority view was a second, rather bold idea. According to this camp, nebulae were not gaseous clouds but entire star systems in their own right, just like the Milky Way. Today we would call them "galaxies"; but at that time they were known as "island universes," a term coined by Kant, who first proposed the idea.

To readers of this book the idea of *other* galaxies will no doubt seem mundane, but for people of the eighteenth and nineteenth centuries this was a truly astonishing proposition. In theory, the Newtonian cosmos was infinite, but in practice most astronomers believed the Milky Way was the *totality* of the universe. Instead of the cosmic infinitude of Cusa and Bruno, most scientists had retreated to the idea a single island of stars amid a vast empty void. When Hubble came on the scene in the early 1920s the burning question in astronomy was whether ours was the *only* cosmic island, or whether there were others. Were we alone galactically, or was Kant correct in supposing a multitude of "island universes"?

Like Kant, Hubble suspected that each nebula was indeed an entire galaxy—though he always rejected this modern neologism, preferring instead the old-fashioned Latin term. But nothing could be resolved in the "great nebula debate" until someone found a way to measure the distance to these cosmological blobs. If nebulae were merely clouds of gas *inside* the Milky Way, then they ought to be relatively close; but if they proved to be *outside* our galaxy that would support Kant's hypothesis. As a leading nebulae expert, Hubble felt his powers equal to the thorny problem of ascertaining their distances, and as a staff member at the new Mount Wilson Observatory, he was one of the lucky few astronomers with regular access to the huge new 100-inch telescope—at the time the world's largest.

In the mid-1920s Hubble set out to measure the distances to a number of nebulae, which was, in effect, an exploration of the extent of the universe as a whole. What really was the *scale* of our cosmos? It was a monumental question, and it was being asked by a man with a monumental will to succeed. The key to the strategy Hubble devised for measuring nebulae distances was a discovery made some years earlier by the pioneering woman astronomer Henrietta Leavitt. Leavitt had found that a certain kind of star known as a Cephid variable could be used as an interstellar measuring stick. Cephids have the unusual property that they periodically pulse, getting brighter and darker in a regular cycle that lasts anywhere from several hours to several months. Leavitt determined that the longer the period, the brighter the star would shine. This distinctive period-brightness relationship meant that Cephids could be used as "beacons for calculating distances across the void."[5] That is, they functioned as a kind of standardized celestial tape measure.[6]

Hubble decided to look for Cephid stars *inside* nebulae, and if he found any, to use these to determine the nebulae's distance from us. It was an inspired move because at the time no one was even sure if nebulae contained *any* stars. If these fuzzy blobs were just clouds of gas (as many astronomers suspected), no stars would be expected. With the huge new Mount Wilson telescope at his disposal Hubble discovered that indeed nebulae *did* have stars, and some even had Cephids. The figures he calculated for their distances were simply staggering. At a time when many astronomers believed the Milky Way (and thus the whole universe) to be no more than thirty thousand light-years across, Hubble calculated that the Andromeda nebula was a million light years away! No one had ever seriously considered such immense distances before; small wonder it was so difficult to see the individual stars. Kant had been right all along; these fuzzy blobs were entire "island universes" each consisting of millions, even billions of stars.

Given the distances revealed by Hubble, the whole scale of the universe suddenly took a quantum leap upward. It is all very well to speak in theory about an infinite universe, but until Hubble's nebulae work few people seriously envisaged a cosmos without end. Now with concrete evidence of *other galaxies* at hand, the old vision of Cusa and Bruno at last began to seem real. With each generation of telescopes, cosmological space has continued to get bigger, for the further out astronomers have looked, the more galaxies they have found. As far as we can tell today there is no end to cosmological space. "In effect," says Robert Romanyshyn, the telescope has "opened, enlarged, and expanded the world," giving us an even greater sense of cosmic enormity.[7]

If Hubble had done nothing else but establish the existence of other galaxies, and hence the true scale of the cosmological whole, he would have gone down in history books, but his greatest achievement still lay ahead. In 1928 he turned his attention to another facet of nebulae, which this time would lead to a totally unexpected conclusion. This time, inspiration came from fellow astronomer Vesto Slipher, who in 1914 had discovered what is known as the cosmological "redshift." Every star, like every household lamp, has a spectrum of light that it emits. Now, just as a train whistle changes pitch as it hurtles past you, getting lower as it moves away, so the "pitch" or frequency of a moving *light* will also get "lower" when it speeds away from you. With light, the result is that it will appear *redder* than it actually is. Slipher had discovered that the light spectrums of some nebulae were considerably redder than the norm—hence he realized that they must be speeding away from the earth.

During the time that nebulae were thought to be just clouds of gas, Slipher's discovery seemed little more than a curiosity, but with Hubble's news that nebulae were in fact galaxies, these redshifts acquired new significance. That something as large as a galaxy containing millions of stars could be moving at all seemed extraordinary. But what could it mean that entire galaxies were hurtling through space at enormous velocities? Again Hubble had an inspired hunch. He imagined that the *further away* a nebula was, the *faster* it might be moving, and hence the *more* its spectrum might be redshifted. He had no particular reason for thinking this; it was just his astronomer's nose twitching in the cosmic wind. Whatever the reason, it was a brilliant imaginative leap, evidence that science does not proceed by logic alone.

Hubble threw himself into this new problem, this time assisted by Mount Wilson's undisputed technical king, Milton Humason—an unschooled man with the rare distinction of having risen to the rank of astronomer after having started at the telescope as a janitor. This "unlikely pair," the Rhodes scholar and the janitor, set to work measuring galactic redshifts.[8] Within months Hubble's hunch had been confirmed; the further away a nebula was, the greater its redshift, and hence the faster it was moving. When Hubble plotted a graph of distance against redshift, the result was a crisp straight line, a beautiful linear relationship. Hubble was ecstatic at this discovery, for he had found a new mathematical harmony in the stars. More importantly, beneath what at first appeared to be a rather esoteric technical finding lay a cosmological bombshell.

Like the edge of an executioner's blade, the line of Hubble's graph cut clean through the static Newtonian cosmos. As Ferris explains: "Inscribed in Hubble's diagram was the signature of cosmic expansion."[9] In other words, what Hubble's graph revealed was the astonishing fact that the *universe is expanding!* His distance-redshift relation implied that not only are the galaxies rushing away from our earth, they are all rushing away from *each other*—which meant that the entire universe must be getting bigger. With each passing second, every galaxy gets further away from every other one, like shards of some monstrous explosion. The entire galactic network is exploding with phenomenal energy, each minute expanding the size of the visible universe by billions of cubic light years. With Hubble's innocent-looking graph, the notion of a static, timeless universe was thereby shattered. Suddenly the cosmic whole was seen to be *dynamic*. Ironically, Hubble—who was not generally a man to shun glory—never quite reconciled himself to this obvious interpretation of his work. "Years later, he was still describing the notion of cosmic expansion as 'rather startling.' "[10]

More startling still was what this cosmic expansion implied—and herein lies the real revolution of twentieth-century cosmology. If all the galaxies are rushing away from each other, making the universe even *larger*, then logic dictates that in the past the universe must have been *smaller*. Playing the cosmological tape backwards, there must have been a time when galaxies were *not* separated by the vast distances we see today, but were packed in close together. From the evidence of cosmic expansion thus came the conclusion that the universe had a *beginning*, a small dense phase out of which the vast modern cosmos exploded. Commenting on this scenario in a BBC radio interview, the English astronomer Fred Hoyle coined the pithy term "big bang." Hoyle used the term in a derogatory manner—he thought the whole idea was rubbish—but the moniker stuck, and today it is one of the most famous phrases in science.

With galactic expansion and the big bang, physicists had unexpectedly stumbled upon their own sense of cosmic *history*. Not just the armchair theorizing of Kant and Laplace, but an empirically grounded basis for a cosmological story. Here at last was the start of a purely physical narrative of creation: the first step on the road to a scientific account of cosmological unfolding. No one could have been more surprised than Hubble, who to the end of his days remained uncomfortable with the whole idea.

Hubble and Hoyle were by no means the only astronomers uncomfortable with the idea of a big bang. Initially, *many* scientists hated the idea because it seemed to smack of religion. If the universe had a beginning, they thought, then it must have had a creator—but that would be unscientific. Yet amazingly, this "heretical" idea was supported by a bold new theory at the very forefront of scientific thinking. It was not a theory familiar to Hubble, or to most astronomers of the 1920s, but soon the whole world would hear about it. The architect of this theory, a young German physicist named Albert Einstein, had himself been so swayed by the tradition of cosmic stasis that even he was initially unable to accept the idea of a cosmic origin. Thus, although his theory predicted just that, Einstein chose instead to fudge his equations to get rid of the cosmic motion they implied. For once in his life, this great iconoclast lost his nerve. By fudging his equations Einstein missed the opportunity to make what surely would have been the most spectacular prediction in the history of science. In doing so he also marred the pristine beauty of that singularly awesome scientific jewel, the "general theory of relativity."

Einstein had actually completed his theory in 1916, more than a decade before Hubble's discovery. From early on, he and others had realized that the relativistic equations predicted a nonstatic universe. Yet on questioning astronomers at the time Einstein was told there was no evidence for this. As a general rule Einstein had no qualms about putting his beloved theories before contradictory observations, but this time he balked. Conceding to the prevailing view, he corrupted the symmetry of his equations to force cosmic stasis. He later called this "the greates blunder of my life." Unlike Hubble, however, once galactic motion *was* discovered, Einstein embraced the expanding universe and all it implied. If Hubble made the discovery of a dynamic cosmos, it was Einstein's equations that made sense of this extraordinary finding. Encoded in the theory of general relativity was a mathematical story of how a universe could unfold out of nothing. Here, in the language of geometry, was a rigorous account of cosmic creation, a scientific rival to the six days of Genesis.

What is most surprising here is that general relativity had not been designed to answer cosmological questions. Einstein's original concern was not with the architecture of the cosmos but with the everyday laws of physics—the same laws that had so exercised the young Newton. Yet in trying to reconcile anomalies in these basic physical laws Einsten had been led to a new conception of space; it was from this new understanding of space that the vision of an expanding universe had unexpectedly emerged. Once again, then, we see that a new conception of space would entrain a new vision of the cosmological whole.

Einstein's conception of space is truly radical. Not in their wildest dreams could Newton and his followers have imagined what a complex, multifaceted entity space would turn out to be. This new vision catapulted the wild-haired German into the stratosphere of celebrity, and it is a measure of the central role of space in the contemporary world picture that while few people understand what Einstein's theories mean, he has become one of the premier icons of our time. His story has been told so many times (including by me),[11] that it is has become difficult not to sound tired when recalling his life. But since science is always a personal, as well as a social, project, it is illuminating to know something of the psychological force behind the work—and in Einstein's case we are dealing with a particularly powerful psyche. For all the mythology of a shy bumbler, the real-life Einstein was a force to be reckoned with.

So much has been made in the Einstein mythology of his dismal record at school that he has become almost the patron saint of scholastic failure. What is less well-known is that here was a child who at age twelve picked up a book and taught himself geometry. Einstein later called this text "the holy geometry booklet," and the deep impression it made on him would resonate throughout his life, for general relativity is above all a *geometric theory of space*.[12] All his life Einstein eschewed the normal scholstic path and turned his attention to the things that interested him. This unorthodox bent, allied with a healthy sense of his own ego, combined to make Einstein a character unlikely to win approval from teachers. Consequently, he had a hard time getting a job after graduation. Finally, as the legend tells it, he landed a job in the Swiss Patent Office as a "technical expert, third class."

Much has been made of the "genius" forced to endure this ignominious position, but Einstein himself always spoke fondly of the patent office and later referred to it as "that secular cloister where I hatched my most beautiful ideas."[13] Here his job was to check patent applications, and across his desk came all manner of mechanisms, including the odd shot at that old favorite, the perpetual motion machine. The patent office experience gave Einstein a deep love and knowledge of machines that once again belies the myth of the absentminded boffin. When in 1931 he was invited to visit Hubble at Mount Wilson, he astonished everyone by climbing all over the telescope and gleefully describing its many parts. His interest in practical machinery also led to a partnership with fellow physicist Leo Szilard to design safer household refrigerators.[14] So much for the man who supposedly couldn't work a can oppener!

But if Einstein apparently enjoyed his job at the patent office, his primary interests undoubtedly lay in more theoretical directions. In between checking patents, the young engineer applied himself to rethinking the foundations of physics. In particular, he thought about the nature of space and time. The first fruit of his innovative mind was not the general theory, but the simpler "special theory of relativity," precursor to the more general variety. It is with this theory that we first glimpse the radical direction in which Einstein would take our conception of space.

The special theory of relativity emerged out of a penetrating critique of Newton's idea of *absolute space*. This is the notion that space forms an absolute backdrop to the universe, an absolute frame against which everything else can be uniquely measured. Despite Newton's tyrannical insistence on this notion, there had always been dissenters, most notably his arch-rival Gottfried Wilhelm Leibniz. From the start, Leibniz had objected that the idea of absolute space was a logical absurdity. Opposing Newton, he suggested that space and time were purely *relative* phenomena. Against this truly perceptive vision, Newton threw the full weight of his authority; and also his theology, which specifically associated absolute space with God. Since Newton believed that God was absolute, he insisted that space also must be. In the bitter debate that ensued, Newton eventually won. "Nothing that Leibniz [and others] had to say in criticism of Newton's concept of absolute space could prevent its acceptance," and for the next two hundred years most physicists just blithely accepted the maestro's view.[15]

Attempts were even made "to demonstrate the logical necessity" of absolute space.[16] Here again we find Kant leading the way. In an effort to shore up the concept of absolute space and time, Kant tried to demonstrate that they were *necessary* aspects of a scientific world picture. "A priori" categories, he called them, and succeeded in convincing himself and a good many others that the entire matter had been resolved. It is a testament to Einstein's enormous self-confidence that while still in his early twenties (without yet even a Ph.D. to his name) he could challenge the collective authority of both Newton *and* Kant. By rejecting the idea of absolute space, the young engineer was pitting himself against the titans of both science *and* philosophy.

In taking this step Einstein was prompted by a dilemma which at that time was occupying some of the finest minds in physics. To these men it was increasingly clear that their science was facing a crisis. The essence of the problem was that the speed of light *always* appeared constant. Why this should be so troubling can be understood by considering not light, but cars. Let us indulge, then, in what Einstein called a "thought experiment." Imagine two cars speeding toward each other on a highway. If one is traveling at fifty miles an hour and the other at forty, then their velocity relative to one another will be ninety miles an hour. In Newtonian physics, as in everyday experience, velocities add together—which is why head-on collisions tend to be fatal.

Now according to the equations that describe the nature of light (Maxwell's equations), the speed of light in empty space is 186,000 miles per second. Quite reasonably, physicists assumed that as with cars, so too with light; velocities would *add* together. Thus, if I were traveling at 1,000 miles per second toward a lamp, the velocity of its light *relative to me* would be 186,000 plus 1,000 and hence 187,000 miles per second. But when two scientists did an experiment to test this assumption, they found to their surprise that regardless of the speed of the observer, light *always* appeared to travel at exactly 186,000 miles per second. No more and no less. Unlike cars, the velocity of light seemed to be the same relative to everything.

This was like Alice in Wonderland physics, and inside the ivory towers of academe the storm clouds began to gather. But down at the patent office Einstein was quietly reinventing the world. Rather than trying to explain away the constant speed of light, Einstein just accepted the fact at face value. Instead of wringing his hands, he asked himself bluntly: How *might* one explain that light travels with the same velocity relative to everyone? If, for example, I travel at a *different* speed to you, how can light appear to travel at the *same* speed relative to us both?

With one of those magnificent intuitive leaps for which he is justly famous, Einstein realized that the problem lay with Newton's insistence that space and time are absolute. He saw that if he abandoned absolute space then the whole problem would disappear. Instead of everyone sharing one universal space and time, he saw that if everyone occupied his or her own private space and time the dilemma would be resolved. In each person's private space the speed of light would be constant for him or her. According to Einstein, then, space and time are not absolute, but purely *relative* phenomena, just as Leibniz had argued two hundred years earlier. Moreover, Einstein was able to put this idea into rigorous mathematical form, showing precisely how space and time would vary according to the *velocity* of each observer. The greater the velocity between two people, the greater would be the difference in their perception of space and time. To sum up: the faster I go relative to you, the more your space will appear to shrink and the more your time will appear to slow down.

The initial reaction to this astonishing proposition was blank disbelief—and needless to say, job offers did not come flooding in to the patent office. To most physicists the idea that space and time could be private affairs seemed utterly ludicrous. Yet special relativity worked. Not only could Einstein successfully explain the constant speed of light, his elegant equations made lots of practical predictions about such concrete phenomena as the behavior of electrons in magnetic fields. Without an understanding of special relativity, for instance, you would not have electric power coming efficiently to your home. In the end, relativity's sheer practical force won over the skeptics. More so since by now it had also become patently "obvious that absolute space evaded all means of experimental detection."[17] Slowly but surely, what physicist Ernst Mach had decried as "the conceptual monstrosity of absolute space" gave way to the liberating vision of *relative space* (and time).[18]

By the age of twenty-six, Einstein had already revolutionized the scientific understanding of space, yet special relativity dealt only with the case of bodies moving at *uniform velocity*. If something as simple as uniform straight line motion could radically alter our experience of space and time, what effect might *nonuniform* motion have on these phenomena? In other words, what would be the effect of *acceleration* on space and time? Even while he was still at the patent office, Einstein had begun to dream about an even grander theory that would encompass the general case of all motion, a theory that would describe what happened to space and time under *all* dynamic conditions.

If before the young engineer had been sailing in uncharted waters, now he was entering territory whose very existence most physicists had never even suspected, the kind of region on medieval maps where one might encounter the warning *Here, there be monsters*. And rarely has a physicist faced such formidable mathematical monsters. Still in the thick of it, he wrote to a friend: "One thing is certain: that in all my life I have never before labored so hard. . . . Compared with this problem, the original theory of relativity is child's play."[19] Miraculously, it turned out that in the previous century a German mathematician named Georg Riemann had developed precisely the tools Einstein now needed. Finally, in 1916, using Riemann's new geometry, Einstein succeeded in generalizing his theory. The fruit of his labor was ten extraordinary equations, one for each year of effort: the general theory of relativity.

General relativity is surely one of science's most esoteric theories, yet we have all been deeply affected by Einstein's masterpiece, for this was the theory that put a time line onto existence itself. By justifying Hubble's discovery, general relativity gave a theoretical foundation to the expanding universe and rooted in the language of mathematics the seminal explosion of the big bang. Here in the cool clear voice of geometry was the affirmation that our universe had a beginning. Moreover, what emerged ineluctably from this theory was that at the moment of the big bang, not only matter, but also space and time were "born."

At the core of general relativity was an even more radical conception of space than in the special theory, a conception that gives rise intrinsically to a *cosmological narrative*. In almost every respect general relativistic space is a thorough departure from the Newtonian past. In the Newtonian picture, space was simply a passive arena in which objects sit. The primary quality of Newtonian space was precisely that it had *no* qualities. Its whole purpose was to serve as a neutral field within which the God-given "laws of nature" could be played out. But the view of space that emerges from general relativity is of something imbued with its own power. In Einstein's picture, space is transformed from a neutral arena to an *active participant* in the great cosmological drama.

In the Newtonian world picture, space was essentially an empty box—three linear dimensions extending forever as a limitless void. By contrast, general relativistic space is like a vast *membrane*. To get a sense of this, physicists often use the analogy of a rubber sheet stretched out like a vast trampoline. Now imagine in your mind's eye that I place a bowling ball on this sheet. As you see, the ball will cause the rubber around it to become distorted, making a depression in the previously flat plane. According to general relativity, this is what a massive body like our sun does to the "membrane" of space. As in Figure 4.1 it distorts the space around itself, causing a "depression" in an otherwise "flat" field. The metaphor is elegant, but the consequences are extraordinary.

Imagine that I now take a billiard ball and hit it in the direction of the bowling ball, not aiming at it directly but slightly off to the side. As the billiard approaches the bowling ball it will move into the region in which the rubber sheet is deformed, and as it does it will be deflected from its original path, being drawn into the depression toward the bowling ball. If the depression is deep

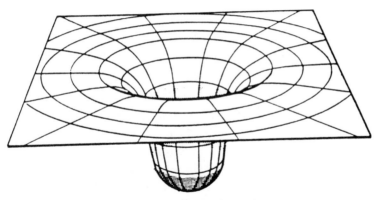

FIGURE 4.1. The sun distorting the "fabric" of spacetime.

enough it will spiral down and come to rest at the bowling ball. This, says Einstein, is the explanation for *gravity*. Rather than being a separate force, gravity is merely a by-product of the *shape of space itself!*

According to Einstein's equations, the more massive a body, the deeper will be the "depression" it creates in space, and hence the greater will be the force of gravity experienced in its vicinity. Physicists refer to this distorting of space as its "curvature." In general relativity, then, gravity is just a by-product of *curved space.* Here on earth we are standing in a small depression in the surrounding spatial membrane caused by our planet—we are in a part of space that is just slightly curved. Looking further afield, the earth itself is in a larger depression caused by the presence of our sun, so space near the sun is more curved. According to general relativity, every star in the universe makes its own curved depression in the spatial membrane, which thereby takes on the character of a landscape. In this vision, space is no longer an inert backdrop, it has become a cosmological terrain—a visceral *substrate* to the universe.

Moreover, just as the presence of matter warps space on a

local scale, it also affects the cosmological whole. Perhaps the most startling consequence of general relativity is that the universe has an *overall architecture*. Again we see here a radical departure from the Newtonian picture, where the cosmos was devoid of form. Applied at the largest scale, the equations of general relativity determine the overall structure of cosmological space. Furthermore, according to Einstein's equations this cosmic architecture is *dynamic*, with an almost organic history.

Once again we may turn here to rubber sheet physics for illumination. This time, instead of a flat rubber sheet, imagine the spherical rubber skin of a balloon. See in your mind's eye a very large balloon, and imagine that you live on its surface—this is the *space* of your balloon universe. Note that we are referring here to the skin *only*, not to the air inside the balloon, so your balloon "universe" is in fact two-dimensional. The equations of general relativity describe our universe as rather like the skin of a balloon. The difference is that while a balloon-skin has only two dimensions, our universe "skin" has four: three for space and one for time. In both versions of relativity, space and time are bound together in a *four-dimensional whole*. Here time becomes, in effect, another dimension of space. This four-dimensional complex is known by the single word "spacetime," but physicists often just speak of four-dimensional *space*, subsuming time into the more general concept of space.

Now according to general relativity, just as the amount of matter determines the degree of space warping on a local scale, matter also determines the shaping of the cosmic whole. Put simply, the *more* matter there is in the universe, the more its overall space will be curved. If there is enough matter, the universal space will be a closed surface, like a balloon (only in four dimensions). If there is not enough matter, the universal space will be an "open" form that physicists liken to a saddle. One of the major challenges of late twentieth-century astronomy is to measure the amount of matter in our universe and thereby determine the specific architecture of the cosmological whole.

Whether the universal space is closed in on itself like a balloon, or open like a saddle, it is definitely expanding. Whatever its shape, general relativity tells us that our universe has a built-in propensity to swell. To understand what this means, imagine yourself back on the balloon-skin, and imagine this time that someone has drawn on the balloon's surface a random assortment of dots. Each dot represents a galaxy. Now, in your mind's eye, imagine that someone is blowing up the balloon. As it expands what you would see is that all the dots (that is, all the galaxies) would appear to move away from you. Moreover, the *further* away a spot was, the *faster* it would appear to be speeding away. In other words, the balloon-skin dots would behave just like Hubble's galaxies. There is nothing magical going on here; it is a simple fact of geometry on an expanding sphere that each part of the surface will move away from each other part at speeds proportional to the distance between them.[20] Hubble's redshift relationship is simply a reflection of this underlying cosmic expansion.

According to general relativity, our universe is behaving like a four-dimensional expanding balloon. It is the *space itself* that is expanding, like a balloon skin. The galaxies of our universe are not hurtling away from one another into an *already existing* space; rather as the space itself expands its reach, it takes the galaxies with it. Space, in a sense, becomes like a living thing—a continually swelling cosmic fruit. The scale of this cosmic expansion is truly staggering. "Every day," says physicist Paul Davies, "the region of the universe accessible to our telescopes swells by 10^{18} cubic light-years."[21] That is, a billion billion cubic light years every single day!

The inherent dynamism of general relativistic space encodes within it the story of its genesis. Since this space is constantly getting bigger, logic dictates that in the past it must have been smaller. Extrapolating back, the whole of cosmological space that we see today must once have been confined to a very small region, to a microscopic point in fact. This infinitesimal speck, which Einstein's equations precisely predict, is the initial spark of the big bang.

Yet despite Einstein's equations, and Hubble's observational evidence, many scientists were at first bitterly opposed to the idea of a cosmic origin, which seemed to raise the specter of a Christian-style Creator—that awkward issue physicists had been avoiding for two hundred years. So reluctant were some physicists to align their science with anything that even hinted of Christianity, several ingenious theories were advanced to explain how the universe might be expanding *without* there being an initial cosmic moment. Not until the 1970s could anyone definitively prove that there *must* have been a big bang. Stephen Hawking, together with his Oxford mentor Roger Penrose, finally demonstrated, using general relativity, that in a universe such as ours there must have been an initial moment of cosmic coalescence. Freighted though it may have been with the "stigma" of religion, cosmic creation was here to stay. Calculations by astrophysicists now place this event at between ten and fifteen billion years ago.

In the past half century scientists have not only discovered a cosmic beginning, they have developed an entire story of cosmic evolution. Starting from the first speck of creation—the point at which space and time came into being—they have articulated a process by which they believe our universe has unfolded into being. Moving on from the big bang, astrophysicists have pieced together an account of the formation of galaxies, stars, and planets. In parallel they have discovered processes by which stars synthesize in their interiors the chain of atomic elements. If the big bang gave rise to the basic particles—the protons, neutrons, and electrons—it is the stars that have given us the atoms of our flesh and bone, the carbon, nitrogen, oxygen, and so on.

From stasis to story, science has at last articulated its own cosmological narrative. Where Darwin's theory of evolution challenged the biblical account of the creation of life, so relativistic cosmology challenges the Genesis story of cosmic creation. Crucial to this new story is the new relativistic vision of space. Just as the Aristotelian cosmos of the late medievals mirrored the Aristotelian conception of space, and as the Newtonian cosmos mirrored the Newtonian conception of space, so also the Einsteinian cosmos mirrors the Einsteinian conception of space. Here, both are seen as structured and dynamic. Once again, then, space, that most seemingly ephemeral entity, grounds and determines our cosmological scheme.

From Aristotle to Einstein a truly revolutionary shift has occurred in our conception of space. For Aristotle, space was but a minor and rather unimportant category of reality. Newton, by contrast, made space the formal background of his universe, the absolute frame of all action. Yet Newtonian space possessed no intrinsic qualities of its own, being just a formless and featureless void. As such, says physicist Andre Linde, in the Newtonian scheme space "continued to play a secondary, subservient role," serving merely "as a backdrop" for the action of matter.[22] With general relativity, however, space becomes for the first time a *primary active* category of reality. According to relativity, you *cannot* have material objects without a supporting membrane of space. Space thus becomes in Einstein's vision a major pillar of the modern scientific world picture.

The foundational nature of space in the relativistic world picture lends to this previously passive and rather dull entity nothing short of its own personality. Instead of being just an empty arena, space becomes an active participant in the cosmological drama, a visceral entity imbued with its own powers. Moreover, because space is a membrane shaped by matter, when the distribution of matter *changes* so does the landscape of space. For instance, when a star ends its life in the explosion of a supernova, relativity tells us that it sends out great waves of gravity. But since gravity is "only the warping of spacetime," then gravity waves are actually *waves in the membrane of space.*[23]

Similarly, the motion of galaxies also alters the landscape of space. Aside from their general expansive motion, many galaxies are also moving *across* the universe in great cosmic currents. All this motion is reflected in the local warping of space, which like a geological landscape shifts and heaves over eons of time. Littered throughout the universe, physicists also believe, are vast "cosmic strings," and "sheets," lines and planes millions of miles long concentrating vast gravitational power that also dynamically warp the structure of space on an intergalactic scale.

Because of its inherent dynamism, the relativistic membrane of space might perhaps be more aptly compared to a seascape rather than a landscape. Like the terrestrial ocean, relativistic space is wracked by waves, currents, and vortices—a vast fluid four-dimensional surface seething and rippling like an interstellar sea. Upon this relativistic ocean has been launched the armada of the twentieth-century science fiction imagination. Leading the fray to "boldly go where no man has gone before" is the starship *Enterprise.* In its warp drive mode (which enables faster-than-light travel), the *Enterprise* sails on a wave of space, its engines forcing the membrane of space to expand behind the ship and contract in front. At least, that's the scenario a real physicist has suggested as a solution to the general relativity equations "which would correspond with 'warp travel.' "[24]

Star Trek's fictional Starfleet Command are not the only ones looking to harness the fluidity of relativistic space. Real-life Starfleet Command, aka NASA, already has a team investigating new kinds of space propulsion, including those based on relativistic space warping. Known as the Breakthrough Propulsion Physics (BPP) steering group, the NASA team's aim is to transcend the limits of rocket power. According to the group's cofounder Al Holt, general relativity provides a key to this goal. As one of his colleagues opines, "warp drives have some degree of validity."[25]

Relativistic space surfing will not all be smooth sailing, however. As with terrestrial oceans, the dynamism of relativistic space can constitute a significant danger. As any mariner knows, one must beware the face of an angry sea. Most mythic of the dynamic faces of relativistic space is the *black hole*. Made famous by Stephen Hawking (though first suggested by English mathematician John Michell in 1783, and named by American physicist John Wheeler in 1967), black holes are such deep depressions in the relativistic spatial membrane that nothing which falls in can ever escape — not even light.

Inside a black hole, space is so deeply distorted (so curved) that anything crossing the threshold — known as the "event horizon" — is sucked into the maw below and eviscerated. To cite Hawking's blunt appraisal: "If you jump into a black hole, you will get torn apart and crushed out of existence."[26] But since gravity is, after all, just a by-product of the shape of space, the fate that awaits inside a black hole is to be torn apart by space itself. Convulsed and distorted beyond endurance, space around a black hole wreaks its revenge on matter like a cosmological dragon, gobbling up everything that strays too close to its lair. So powerful is the maw of a black hole that space here could rip apart a spaceship. This is how far we have come from the passive picture of Newton — in the relativistic vision, space has become literally *monstrous*.

Yet as with many mythic monsters, if one can evade the teeth of a black hole, great rewards potentially lie on the other side. Here, the prize is access to its mirror image, a *white hole*. As Hawking explains, white holes arise from the fact that "the laws of physics are time-symmetric." Thus, "if there are objects called black holes into which things can fall but not get out, there ought to be other objects that things can come out of but not fall into."[27] Such white holes, according to Einstein's equations, are connected to black holes by tubelike tunnels of space known as *wormholes* (see Figure 4.2). These wormholes have long appealed to science fiction writers as potential engines of space travel.

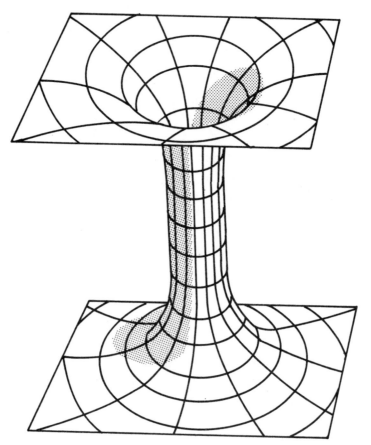

FIGURE 4.2. A wormhole in spacetime.

Instead of traversing billions of miles of space, why not just dive into a black hole, slip through its wormhole, and reappear at your desired destination via the terminating white hole? Through a wormhole, one could in theory tunnel through space like some intergalactic mole (see Figure 4.3). This is indeed the scenario envisioned in the *Star Trek* spin-off series *Deep Space Nine* in an episode centering on the *Bajoran* wormhole. An even more remarkable wormhole appears in the *Voyager* series, and enables the *Enterprise* to travel through time as well as space. This is no mere science fiction fantasy, for as physicist Lawrence Krauss notes, "If wormholes exist, they can and will be time machines" — time, in general relativity, being just another dimension of space.[28]

According to Hawking, "there *are* solutions of Einstein's general theory of relativity in which it is possible to fall into a black hole and come out of a white hole."[29] Unfortunately, these solutions are so unstable that "the slightest disturbance, such as the presence of a spaceship, would destroy the 'wormhole.' " Needless to say, that has not stopped science fiction writers from continuing to dream. Nor has it stopped physicists. An enterprising team at the University of Newcastle in England, for example, has found that if you had two black holes that were electrically charged, a stable wormhole could form between them.

But busy star travelers will surely not rely on naturally forming wormholes, which may be rather rare; they will want to *construct* their own wormholes to specific destinations. Considering this problem while writing his novel *Contact*, astronomer Carl Sagan turned for advice to Caltech relativity physicist Kip Thorne. Sagan's query prompted this distinguished theorist to devise a scheme in which traversable wormholes could be produced by an exotic kind of antigravity-inducing matter containing something physicists call "negative energy." According to Thorne, this is a scenario that general relativity *would* allow.[30]

What is extraordinary here is that under general relativity,

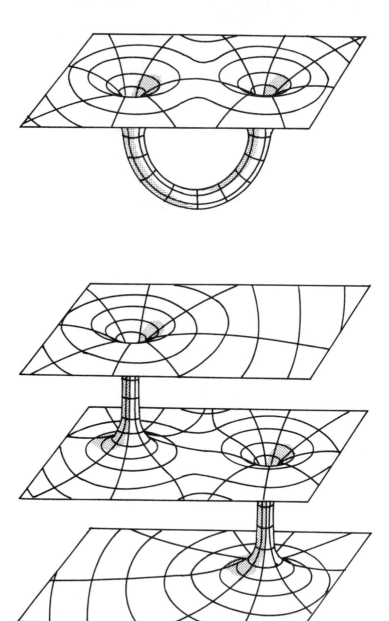

FIGURE 4.3. Wormholes potentially allow travel across spacetime, and even between the different spacetimes of different universes.

space has become not just a sea on which matter might sail, but a highly malleable *substrate* capable of forming complex *structures*. One of the more bizarre spatial structures that relativity physicists now speculate about are "baby universes"—little bubbles of space-time that bud off from our mother universe via black holes. According to the latest theories, our universe is "surrounded" by a foam of these baby universes constantly budding off, then joining back on. Each baby universe has its own unique microscopic spacetime. Even at the macroscopic level, some physicists believe that ours is not the only spacetime. For Andrei Linde and Penn State theoretician Lee Smolin, our universe is but one of a po-tentially infinite array of universes. As in Figure 4.4, each one of these universes is a vast spacetime bubble in its own right. In Linde and Smolin's vision, there is thus a universe of universes, a super-space of cosmological spaces.[31] Some physicists believe we may even be able to get to these other universes by tunneling through wormholes.

The dynamic vision of space described by general relativity is of great importance for professional astronomers and cosmolo-gists, but for most nonscientists surely its primary appeal is the possibilities it seems to engender for extraterrestrial contact. Ever since Kepler's lunar lizards, people have speculated about the fab-ulous other beings we might find among the stars. Space may now be endowed with a character of its own, but it is the characters who might *inhabit* this space that fuel modern cosmological dreams. The great space epics—Isaac Asimov's *Foundation* series, Frank Herbert's *Dune*, and George Lucas' celluloid *Star Wars* trilogy—all derive their enduring appeal from their rich tapestry of inter-stellar cultures and alien mind-sets. In these classics, space is not so much a "final frontier" as a psychological seedbed in which the writers plant the exotic fruits of alternative ways of being.

The allure of alien life, and the tremendous desire for ex-traterrestrial contact, was evident in August 1996 when NASA sci-

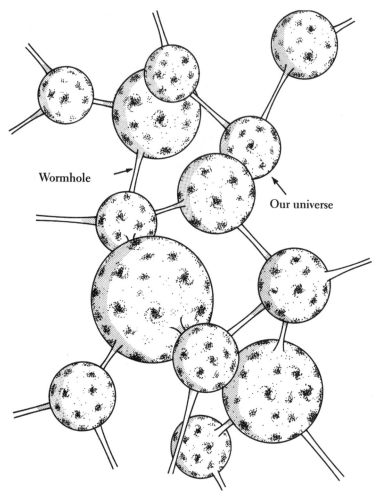

Wormhole

Our universe

FIGURE 4.4. Our universe may be just one of an infinite number of parallel universes, each connected to others by wormholes.

entists announced they had found evidence of fossilized microbes in a meteorite from Mars. The world's press exploded in a frenzy of excitement: Here at last seemed proof that we were not alone. For NASA, of course, evidence of extraterrestrial life—even if only in fossil form—would have been almost as good as a phone call from ET himself. What better way to reenergize public enthusiasm—and funding—for its moribund space program? The enduring appeal of both *Star Trek* and *Star Wars* speaks to the immense psychological yearning for extraterrestrial encounters. As also, in a perverse way, do the thousands of supposed alien abductions and the vast webs of conspiracy theory woven around such mythical alien "landing sites" as Roswell.

Yet as a potential realm of friendship, outer space has so far proved profoundly disappointing. To date astronomers have found no concrete evidence for extraterrestrial life. Men have walked on the moon, and as we all know, they encountered no giant lizards. Visits by space probes to Venus, Mars, and Jupiter have likewise revealed sterile lifeless environments. After two decades of ruthless silence from the stars, in 1993 NASA canceled its Search for Extraterrestrial Intelligence project. (Although the effort has since been picked up by private interests.) Even the Martian microbes turned out to be just residues of geochemical processes. Then, as now, space remains eerily silent.

Our failure so far to locate any extraterrestrial companions is not the only unnerving aspect of contemporary cosmology. Not only are we for all practical purposes alone in our vast relativistic space, we are also in a profound sense *nowhere*. The very "cosmological principle" that Nicholas of Cusa introduced to enhance humanity's cosmic status has in the end backfired against us. If at first it seemed a cosmic promotion to make all places in the universe equal, this democratizing strategy has ultimately stripped us of all cosmic significance.

On the level playing field of the new cosmology we are nowhere special because the very definition of relativistic space guarantees that there is *no place special* to be. Unlike the medieval cosmos in which every place had an intrinsic value (depending on its proximity to God), the equations of general relativity encode *no* sense of value. Einsteinian space may be geometrically precise, but it is also value-free. On the mathematical membrane of general relativity, any place is as good, or as bad, as any other, and it matters nothing whether we humans are here or there or anywhere else. In the limitless depths of the new cosmological space, our earth becomes just an insignificant planet, revolving around an insignificant star in an insignificant galaxy, which in a map of the cosmic whole is lost from view.

We find ourselves, then, in a paradoxical situation, for while we are the first culture in human history to have a detailed map of the entire physical cosmos, we are, in effect, *lost in space*. All those "island universes" seen through our telescopes serve only to reinforce what a puny and insignificant island we truly are. Striving for self-respect in this immeasurable ocean of space, is it any wonder we have turned to the stars seeking friendship and meaning? Is it any wonder that we long to be part of some intergalactic community imbued with purpose and direction?

But herein lies the final indignity of modern cosmological space, for as long as everywhere is *equal* then it matters not whether you seek in one direction or another. In a space where all places are essentially the same, so too are all directions. The mathematization of space has turned the compass into a roulette wheel: Any direction might lead to somewhere exciting. On the other hand, it might not. Here then is a major difference between *The Divine Comedy* and *Star Trek*. While both are cosmological journeys that use their otherworldly peregrinations for reflection on the human condition, Dante's journey was intrinsically *directed*.

The very space Dante traversed encoded the direction he must travel—upward, toward God. His was not a homogeneous realm, but a spiritually graded domain in which the value of each place was visibly evident in the quality of its environment. Dante's compass was always and ever the vector of Christian spiritual improvement, graphically symbolized in the edifice of Mount

Purgatory. Dante did not have a *choice* as to what direction to take: His journey was strictly linear—toward light, hope, and love. Indeed, for the first two canticles, he and Virgil follow an actual path, marked out on the ground before them. Only in the *Paradiso* does the path disappear, but by then Dante's soul is irrevocably propelled toward its target by a force no human could resist.

The *Enterprise*, on the other hand, may go in any direction its captains choose: One region of space has as many dramatic possibilities as any other. Precisely because the *Star Trek* cosmos— that is, *our* scientific cosmos—has no intrinsic directionality, there can be no definitive end to the *Enterprise*'s story. Even if the ship is destroyed, the producers can always commission another and crew it with a new cast of characters. There have already been three separate "generations" of ship and crew. In an intrinsically directed and finite space like that of *The Divine Comedy*, the narrative must end, the goal must be reached, sooner or later. But in an infinite homogeneous space the story can go on forever, which is why *The Divine Comedy* has just three parts, while the *Star Trek* saga is still going strong after more than three hundred episodes.[32]

In a homogeneous space, the traveler has infinite freedom of choice: He can go in any direction he chooses and change his mind whenever he likes. This sense of freedom is a huge part of the fantasy of outer space. It is the same freedom the modern driver feels when cruising the endless highways of America—only in outer space you have three dimensions of movement, four if you also count time. This apparently limitless freedom of movement is a prime fantasy of late twentieth-century cosmology. Yet while we in the West have been developing an ever more detailed and adventure-filled vision of our *physical* cosmos, we have negated the very idea of *other* planes of reality, other "spaces" of being. By homogenizing space and reducing "place" to a strict mathematical formalism, we have robbed our universe of *meaning* and taken away any sense of intrinsic directionality. The flip side of our cosmological democracy is thus an existential anarchy: With no place more special than any other, there is no place ultimately to aim for—no goal, no destination, no end. The cosmological principle that once rescued us from the gutter of the universe has left us, in the final analysis, with *no place to go.*

HYPERSPACE

Clearly . . . any real body must have extension in *four* dimensions: it must have Length, Breadth, Thickness, and Duration. But through a natural infirmity of the flesh, which I will explain to you in a moment, we incline to overlook this fact. There are really four dimensions, three which we call the three planes of Space, and a fourth, Time.[1]

Not unreasonably, one might imagine this encapsulation of the idea of four-dimensional spacetime to be a quote from Einstein. Yet it is not from any physicist; it was written in 1895, fully a decade before the first paper on special relativity, by the science fiction writer H. G. Wells. The statement is from the opening pages of Wells' classic novel, *The Time Machine*, wherein the hero of the story explains to his friends the concept of the fourth dimension and the possibility of time travel. At a time when Einstein was still at school dreaming about riding on light beams, Wells in his fiction was already exploring the consequences of a fourth dimension. In addition to *The Time Machine*, characters in

The Wonderful Visit, "The Plattner Story," and "The Remarkable Case of Davidson's Eyes" all venture into a mysterious extra dimension, there to encounter phenomena impossible in the everyday space of our experience.

Wells was by no means alone among late nineteenth- and early twentieth-century writers in his invocation of other dimensions. "The list of prominent figures" interested in the subject includes Fyodor Dostoevsky, who referred to higher dimensions in *The Brothers Karamazov;* Joseph Conrad and Ford Madox Ford, whose novel *The Inheritors* focused on a cruel race from the fourth dimension; and Oscar Wilde, who made this dimension the butt of his wit in *The Canterville Ghost.*[2]

Artists too were inspired by the notion of a "higher" dimension. Long before relativity filtered into public consciousness, Cubist theoretical writings abounded with references to a fourth dimension, as did the writing of the Russian Futurists. Marcel Duchamp, Kasimir Malevich, and the American painter Max Weber—to name just a few—all went through periods of intense interest in higher dimensional space. So did the composers Aleksandr Scriabin and George Antheil. The fourth dimension also provided impetus to philosophers and mystics. As art historian Linda Dalrymple Henderson has noted, in the late nineteenth century "the 'fourth dimension' gave rise to entire idealist and even mystical philosophical systems."[3] In fact, Henderson says, by the year 1900 "the fourth dimension had become almost a household word . . . Ranging from an ideal Platonic or Kantian reality—or even Heaven—to the answer to all the problems puzzling contemporary science, the fourth dimension could be all things to all people."[4]

Although Einstein's name is the one now most often associated with the idea of a fourth dimension, the concept originally emerged in the mid-nineteenth century. The key impetus was the development of non-Euclidian geometry. From the 1860s on, interest in this new geometry rapidly effervesced into a public fascination with higher-than-three-dimensional space—what came to be called "hyperspace." First explored by writers, artists, and mystically inclined philosophers, this seemingly fantastical concept would eventually give rise to an extraordinary new scientific vision of reality, one in which space itself would come to be seen as the ultimate substrate of all existence. Here, we are not just talking about the extra dimension of time, but also about extra *spatial* dimensions. In this chapter we explore the bizarre story of higher-dimensional space, from its humble beginnings in the mathematics of the nineteenth century to its culmination today with physicists' vision of an *eleven-dimensional* universe.

The bizarre potential of higher dimensional space was evident from the beginning. As early as the 1860s, the great mathematical genius Carl Friedrich Gauss (founder of the new geometry) had begun to think about spaces with four or more dimensions. Significantly, Gauss specifically speculated about the possibility of higher-dimensional beings. Since one cannot imagine a greater-than-three-dimensional world directly, Gauss used an analogy of beings in a two-dimensional world. Here, he envisaged beings "like infinitely attenuated book-worms in an infinitely thin sheet of paper," creatures that would possess only the experience of two-dimensional space.[5] Now just as we can imagine such beings in a *lesser*-dimensional space than our own, so Gauss suggested that we might also imagine beings living in a "space of four or a *greater* number of dimensions." What would such a space be like? What would be its properties? What would it be like to live there? Gauss wondered. Here were the seeds of a science fiction writer's dream—and sure enough, before long the literary responses came pouring in.

One of the earliest and most charming visions of higher-dimensional space was penned in 1884 by the Englishman Edwin Abbott. The theme of Abbott's tale is immediately signaled by its wonderful title, *Flatland: A Romance of Many Dimensions by A. Square*. As the subtitle suggests, the hero of Abbott's adventure is a Square, a being who lives in a two-dimensional space known as "Flatland." In the planar universe of Flatland, a rigid hierarchy reigns. Females, the lowliest beings, are mere straight lines. Males, on the other hand, are regular polygons: squares, hexagons, octagons, and so on. Among males, the more sides one possesses, the higher one's social status. With only fours sides, squares rank at the bottom of the pecking order. Circles, who are infinitely-sided polygons, stand at the top—they are the priests of Flatland. Within this two-dimensional world it is forbidden to talk about, or even to think about, a third dimension, for the idea of anything "higher" than a circle is heresy.

On the plane of Flatland our humble quadrilateral hero is minding his own business, when one night the quiet tenor of his life is shattered by the visitation of a being from the "Land of Three Dimensions." This magnificent creature is none other than a Sphere, a three-dimensional circle! Even in his own world, this paragon of perfection is a lord among his people. In order to demonstrate the inconceivable wonder of the third dimension to the astonished Square, Lord Sphere lifts him up into this higher-dimensional world to see for himself. What especially takes the Square's breath away is the glorious sight of the Cubes he finds there: three-dimensional versions of his own lowly form. So taken is the Square with the expansion of vision he encounters in the third dimension that he urges Lord Sphere onward and upward to higher dimensions still.

> "Take me to that blessed Region where . . . before my ravished eye a Cube, moving in some altogether new direction . . . shall create a still more prefect perfection than himself. . . . And once there, shall we stay our upward course? In that blessed region of Four Dimensions, shall we

linger on the threshold of the Fifth, and not enter therein? Ah, no! Let us rather resolve that our ambition shall soar with our corporeal ascent. Then, yielding to our intellectual onset, the gates of the Sixth Dimension shall fly open, after that a Seventh, and then an Eighth . . ."[6]

Sadly, this "ascent" into higher-dimensional space is not to be, for Lord Sphere is as adamantly opposed to the idea of a fourth dimension as the Circles of Flatland are set against the third. In indignation the Sphere flings the Square back to his two-dimensional world, where he is soon imprisoned for his heretical stories of a third dimension.

If Abbott's Square was unable to reach the fourth dimension, other fictional characters had better luck. In *The Time Machine* H. G. Wells had equated the fourth dimension with time, but in other stories he followed Abbott's example and imagined it as an extra dimension of space. Just as a two-dimensional napkin can be folded within three-dimensional space by bringing together two distant corners, so too within a four-dimensional space two parts of three-dimensional space can be "folded" together. This "folding" of space was the device Wells used in his story "The Remarkable Case of Davidson's Eyes." By judicious folding within four-dimensional space, the hero Davidson is brought into contact with a faraway South Sea island, which he is now able to observe while sitting at home in London. In another of Wells' forays into higher-dimensional space, science teacher Gottfried Plattner is blown away by an explosion and returns from the fourth dimension with his body left-right reversed so that his heart is now on the right-hand side, his liver is on the left, and so on.[7]

For many writers, the fourth dimension would become a place of liberation and redemption, one with distinctly heavenly overtones. Such was the vision of Wells' French disciple Gaston de Pawlowski. In Pawlowski's *Voyage to the Country of the Fourth*

Dimension (1912), he served up a ringing moral tale in which the ability to see and comprehend a fourth dimension saves mankind from scientistic hubris. Within the novel, history was divided into three eras. Beginning in the early twentieth century was what Pawlowski called the "Epoch of Leviathan," an age of rampant materialism and positivism. According to the author this era would culminate during the late twentieth century with a "scientific period" full of nameless horrors. Finally, salvation would come when the fourth dimension was revealed, initiating the "epoch of the Golden Bird." In this "idealist renaissance" man would apparently "raise himself forever above the vulgar world" of three dimensions and find himself in a "higher" realm of wisdom and cosmic unity. As Pawlowski explained: "The notion of the fourth dimension opens absolutely new horizons for us. It completes our comprehension of the world; it allows the definitive synthesis of our knowledge to be realized. . . . When one reaches the country of the fourth dimension . . . one finds [one]self blended with the entire universe."[8]

Pawlowski's heavenly vision of the fourth dimension and his belief in its salvific properties would be widely reflected by others in the first decades of our century. A whole brand of what Henderson terms "hyperspace philosophy" would spring up, giving rise to all manner of curious blendings of science and spirituality. Ironically, the same kind of mathematics that Einstein would later use in the general theory of relativity has also served as a foundation for some of the most bizarre pseudoscientific speculations of our age.

Foremost among the new hyperspace philosophers was Englishman Charles Hinton. As a professional mathematician, Hinton taught at Princeton University and later worked for the United States Naval Observatory and the U.S. Patent Office, but parallel to this orthodox professional life was a mystical underbelly in which he pursued a spiritual approach to the fourth dimension. In *A New Era of Thought* (1888) Hinton outlined a system by which people could supposedly train themselves to become aware of the true four-dimensional nature of space. At the core of this system was a set of special colored blocks, the contemplation of which would supposedly break down restricting "self-elements" within the mind, thereby opening the doors of perception to the fourth dimension.

Hinton dreamed of bringing forth "a complete system of four-dimensional thought—mechanics, science and art,"[9] but in truth he was interested less in the practical applications of the fourth dimension than in its spiritual and philosophical ramifications. Here he was inspired by Plato's analogy of prisoners chained in a cave, doomed forever to see only the shadows of the "real" world outside.

For Hinton, our normal experience of three-dimensional space doomed us to see only the "shadows" of the "real" reality, which is four-dimensional. By becoming aware of this extra dimension, he believed that Plato's realm of the ideal would be revealed. As the realm of the *noumenon*, the fourth dimension could also be seen, in Hintons view, as Kant's "thing-in-itself."

Hinton never realized his "complete system" of four-dimensional thought, but his philosophical interpretation of the fourth dimension would greatly influence later hyperspace thinkers. Among them was the Russian mystic Peter Demianovich Ouspensky. "In the idea of a spatial fourth dimension," says Henderson, "Ouspensky believed he had found an explanation for the 'enigmas of the world,' and with this knowledge he could offer mankind a new truth that would, like the gift of Prometheus, transform human existence."[10]

For Ouspensky, the fourth dimension was none other than time. But according to him, in our everyday experience of this dimension we are deceived. In truth, Ouspensky declared, time is just another dimension of space, and thus all motion is an illusion. According to Ouspensky, the *real* reality is a changeless four-dimensional stasis. Not just time and motion, but matter also is an illusion that people must overcome by learning to "see" anew. Not everyone, however, was mentally equipped for Ouspensky's four-dimensional vision. Those who are so gifted constitute a race of "supermen" with the power to realize what Ouspensky called "cosmic consciousness." In this final state of evolution, the new "supermen" will find themselves graced with "higher emotion, higher intellect, intuition, and mystical wisdom."[11] In this realm, ordinary laws of mathematics and logic will be superseded by a new "logic of ecstasy." It was through just such an "intuitive logic" that Ouspensky proposed to prepare future supermen for the mystical revelation of the fourth dimension.

In Ouspensky's vision of the fourth dimension do we not detect distinct echoes of the medieval Christian Heaven? Just as in the Empyrean time was negated, subsumed into an eternal blissful stasis, so also in Ouspensky's hyperspace realm we find ourselves in a state of ecstatic stasis. Here too in the fourth dimension, we are promised "higher emotion," "higher intellect," even "mystical wisdom." In such early twentieth-century visions of a fourth dimension we witness a recasting into scientific terms the old idea of a transcendent, heavenly domain.

Another hyperspace philosopher with even more overtly Christian leanings was the Rochester, New York, architect Claude Bragdon. It was Bragdon who organized the English translation of Ouspensky's work, and the two men immediately recognized kindred spirits in one another. In addition to Bragdon's more philosophical works, his oeuvre also included a curious little religious tale called *Man the Square: A Higher Space Parable*. Here, Bragdon used the analogy of a two-dimensional world (rather like Abbott's Flatland), "to convey a message of love and harmony."[12]

As in *Flatland,* Bragdon's characters are also simple geo-metric figures living on a flat surface (see Figure 5.1). As the story unfolds, however, we learn that all these figures are really cross sec-tions of cubes, tilted at different angles to their two-dimensional planar world (see Figure 5.2). Seen from the "higher" reality of three dimensions, the beings are *not* flat figures but hearty, solid cubes. At the end of the story, this higher-dimensional reality is demonstrated to the flatlanders by a "Christos cube," which reveals its true cubic nature by folding down its six sides to form the shape of a cross. In the logic of the story, what brings about disharmony in the two-dimensional world is that the cubes of the flatlanders are all tilted at odd angles to their plane. To reinstate harmony, the cubes need to be aligned upright so they are all "square" with their plane. The moral of the tale (of course) was that *we too* need to get ourselves properly aligned in our own higher space dimen-sion—i.e., the fourth.

Along with the supposedly philosophical and moral impli-cations of the fourth dimension, Bragdon was also interested in its aesthetic possibilities. "Consciousness is moving towards the con-quest of a new space," he wrote. "Ornament must indicate this movement of consciousness."[13] To this end, Bragdon produced *Projective Ornament,* a book of images created by projecting four-dimensional figures onto two-dimensional surfaces. The result, as in Figures 5.3 and 5.4, was a kind of geometric Art Deco that was, in truth, rather banal. Bragdon's imagery failed to precipitate the aesthetic revolution he was hoping for, but elsewhere real art-world heavyweights *were* looking to the fourth dimension for in-spiration. And some may even have taken cues from Bragdon's work.

In the canon of modern art, one of the most striking icons is Kasimir Malevich's late 1920s work *Black Square*—a single, stark, black square painted against a white background. Nothing could be simpler; but what does it mean? When asked, Malevich enig-

FIGURE 5.1.
"Personalities: Tracings
of the Individual (Cube)
in a Plane." From *Man
the Square: A Higher
Space Parable* by
Claude Bragdon.

matically responded that it was "a desperate attempt to free art
from the ballast of materiality."[14] Probing further, art historians
have identified a strong link between the Russian Futurists, to
whom Malevich belonged, and the four-dimensional mysticism
then being espoused by their countryman Ouspensky.

In 1913 Malevich had designed the sets for the avant-garde
opera *Victory Over the Sun*, whose writer was consciously trying to

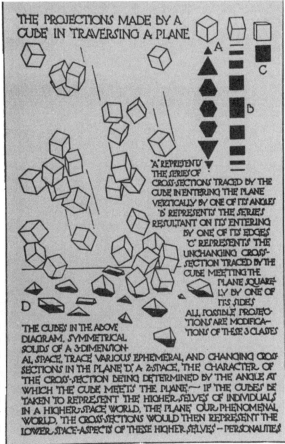

THE PROJECTIONS MADE BY A CUBE IN TRAVERSING A PLANE

'A' REPRESENTS THE SERIES OF CROSS-SECTIONS TRACED BY THE CUBE IN ENTERING THE PLANE VERTICALLY BY ONE OF ITS ANGLES 'B' REPRESENTS THE SERIES RESULTANT ON ITS ENTERING BY ONE OF ITS EDGES 'C' REPRESENTS THE UNCHANGING CROSS-SECTION TRACED BY THE CUBE MEETING THE PLANE SQUARE-LY BY ONE OF ITS SIDES ALL POSSIBLE PROJEC-TIONS ARE MODIFICA-TIONS OF THESE 3 CLASES

THE CUBES IN THE ABOVE DIAGRAM, SYMMETRICAL SOLIDS OF A 3-DIMENSION-AL SPACE, TRACE VARIOUS EPHEMERAL AND CHANGING CROSS SECTIONS IN THE PLANE 'D', A 2-SPACE. THE CHARACTER OF THE CROSS-SECTION BEING DETERMINED BY THE ANGLE AT WHICH THE CUBE MEETS THE PLANE. — IF THE CUBES BE TAKEN TO REPRESENT THE HIGHER SELVES OF INDIVIDUALS IN A HIGHER-SPACE WORLD, THE PLANE OUR PHENOMENAL WORLD, THE CROSS-SECTIONS WOULD THEN REPRESENT THE LOWER SPACE-ASPECTS OF THESE HIGHER SELVES — PERSONALITIES

FIGURE 5.2. "The Projections Made by a Cube in Traversing a Plane." From *A Primer of Higher Space* by Claude Bragdon.

evoke an Ouspenskian new consciousness. In his set designs, Malevich incorporated an image that looks suspiciously like one of Hinton's four-dimensional "hypercubes." At the same time, he had also begun experimenting with the geometric forms that would eventually lead to the new style of Suprematism—of which the *Black Square* is the most famous example. The inspiration for this image may have come directly from Bragdon, who by that stage was in contact with Ouspensky in Russia. According to art

P R O J E C T I V E   O R N A M E N T

the tesseract. The fact that they are not cubes except by convention is owing to the exigencies of representation: in four-dimensional space the cells are perfect cubes, and are correlated into a figure whose four dimensions are all equal.

In order to familiarize ourselves with this, for our purposes the most important of all four-fold figures, let us again consider the manner of its generation, beginning with the point. Let the point A, Figure 8, move to the right, terminating with the point B. Next let the line A B move downward a distance equal to its length, tracing out the square A D.

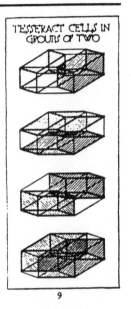

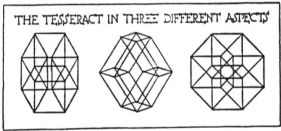

FIGURE 5.3. A page from *Projective Ornament* by Claude Bragdon.

critic, Geoffrey Broadbent, "Malevich's *Black Square* seems to be nothing more, nor less, than his 'Non-Objective' representation of Bragdon's (human-being-as) Cube passing through the 'Plane of Reality'!"[15] Indeed, Malevich had even written a script for an animated film about cubes tumbling through space.

The Russian Futurists' interest in the fourth dimension was inspired by a similar interest among the French Cubists, particu-

FIGURE 5.4.
A plate from
*Projective
Ornament* by
Claude
Bragdon.

larly the writings of the Cubist theorists Albert Gleizes and Jean
Metzinger. Art scholars have heavily debated the link between
Cubism and the contemporary fascination with a fourth dimension, but while it is true that the initial impetus for Cubism came
from other directions (notably from a desire to free painting from
the strictures of perspective), in its later phases many Cubists drew
inspiration from the new non-Euclidian geometry. As Gleizes ex-

plained in a 1912 interview, "beyond the three dimensions of Euclid we have added another, the fourth dimension, which is to say the figuration of space."[16]

That Cubists and other modernist artists should be interested in higher-dimensional space is hardly surprising, for a primary thrust of early twentieth-century art was to break with the tradition of perspective. If it turned out that physical space was *not* in fact three-dimensional, then the rules of linear perspective would simply be arbitrary. The possibility of higher-dimensional space thus served a powerful rhetorical function for the nascent moderns. Recognizing this explicitly, Gleizes and Metzinger stated in *Du Cubism* that "If we wished to tie the painter's space to a particular geometry, we should have to refer it to the non-Euclidean scholars."[17]

The painter who most seriously took up this challenge was Marcel Duchamp. Originally associated with the Cubists, Duchamp soon spun off onto his own peripatetic paths. Like Malevich, his most famous work was also inspired by the fourth dimension. *The Bride Stripped Bare by Her Bachelors, Even*, often known as *The Large Glass*, is surely one of the most pondered-over works in the modern canon; and this time we have extensive notes by the artist detailing the process of genesis. Specifically, we know that in preparing for this work Duchamp embarked on a study of non-Euclidian and higher-dimensional geometries. The end result of these efforts was a complex work divided into two distinct halves: in the top half is the "Bride," and in the bottom half the "Bachelor Apparatus." According to Duchamp's notes the Bride is supposed to be a four-dimensional entity, while the bachelors are three-dimensional. Floating above her retinue, this higher-space spouse hovers enigmatically in a world of her own.

With all this artistic, literary, and mystical speculation about a fourth dimension, what delicious synchronicity when the theory of relativity suddenly enshrined the concept in *physical reality*.

Einstein's revelation of the fourth dimension seemed to many hyperspace enthusiasts a confirmation of what they had known all along. The common thread running between the worlds of relativistic physics and that of the writers and artists was of course the new mathematics of non-Euclidian geometry. Ironically, many of the new-math pioneers had themselves been driven to their radical geometries by a scientific interest in the structure of physical space. To these men, Gauss included, their fantastical new geometries had originally evolved as tools for helping them to better understand the nature of the concrete physical world. Thus while they are generally remembered today as mathematicians, along with Einstein these men ought also to be recognized as pioneers in the physics of space.

In fact, the whole development of non-Euclidian geometry that Gauss initiated emerged out of his work on the measurement of the earth. Given that the literal meaning of the word "geometry" is "earth measurement," this was particularly apt. In its original incarnation, the science of geometry had emerged from ancient Egyptian surveying of the Nile Delta. This ancient (i.e., Euclidian) geometry had only dealt with *flat* space, such as the surface of this page. On a large scale, however, the surface of the earth is spherical, and hence curved. Thus a study of the earth's surface ultimately requires a geometry of *curved surfaces*. Gauss' seminal papers on curved-space geometry were inspired by his stint as scientific advisor to a geodetic survey of the region of Hanover. "Once again," says Max Jammer, "we see that historically viewed, abstract theories of space owe their existence to the practice of geodetic work."[18]

Humans had long known that the surface of our planet is curved, but what about the space in which our globe is *embedded*? Might *space* itself be curved? For Newton and his contemporaries there had been no mathematical alternatives to Euclidian space so they had simply assumed that this was the correct model for physical space. But after Gauss' work on curved surfaces, he began to wonder if the assumption of a Euclidian universe was justified. In the early nineteenth century—long before Einstein was born—Gauss actually tried to measure the curvature of physical space. He did this by the ingenious method of surveying a triangle formed by three mountaintops. In Euclidian, or flat space, three angles of a triangle *must* add up to 180 degrees, but if the space is curved the angles will add up to something else. (To *more* than 180 if the space is "positively" curved, like a sphere, and less than 180 if it is "negatively" curved, like a saddle.) Since Gauss failed to find any deviation from 180 degrees, he concluded that at least in the vicinity of the earth, space must be Euclidian.

The later Russian mathematician, Nikolai Ivanovich Lobachevsky, would try a similar experiment but on a much larger scale. Instead of mountains, Lobachevsky used distant stars, yet still he found no deviation from flat space. Both Gauss and Lobachevsky concluded, based on the evidence available to them, that our local area of the universe was Euclidian, but both realized there was no reason why this *must* be the case. As Gauss presciently put it: "In some future life, perhaps, we may have other ideas about space which, at present, are inaccessible to us."[19]

While Gauss and Lobachevsky pioneered the idea of curved space, later in the nineteenth century a brilliant young mathematician named Bernhard Riemann even considered the possibility that gravity was a by-product of *curvature* in higher-dimensional space. While there is no doubt that Einstein thought up this concept for himself, it is worth noting that the idea had already been imagined more than half a century before. The young man responsible for this astonishing insight was a disciple of Gauss, and he remains one of the most underrated visionaries in modern science. Today Riemann is generally remembered as a pure mathematician, but what really interested this pathologically shy Austrian was the problem of how physical forces arise. Decades before Einstein's birth, Riemann became convinced that the explanation for gravity must lie in the geometry of space.

Thinking about the problem of physical forces, Riemann imagined a world not unlike Abbott's Flatland, in which a race of two-dimensional creatures were living on a flat sheet of paper. Now what would happen, Riemann asked, if we crumpled the paper? Because the creatures' bodies are *embedded* in the paper, they would not be able to see the wrinkles—to them their world would still look perfectly flat. Yet Riemann realized that even if the space *looked* flat, it would no longer *behave* as if it were flat. He argued that when the creatures tried to move about in their two-dimensional world they would feel a mysterious unseen "force" whenever they hit one of the wrinkles, and they would no longer be able to move in straight lines.

Extrapolating this idea to our three-dimensional universe, Riemann imagined that our three-dimensional space was also "crumpled" in an unseen fourth dimension. Like the two-dimensional beings of the paper universe, he reasoned that although we could not see such "wrinkles" in the space around us, we too would experience them as invisible forces. From this brilliant insight, Riemann concluded that gravity was "caused by the crumpling of our three-dimensional universe in the unseen fourth dimension."[20] Having outlined his basic theme, this shy genius set about developing a mathematical language in which to express these ideas. The result of his labors was the new geometry that Einstein would eventually use in his general theory of relativity. "In retrospect," says physicist Michio Kaku, "we now see how close Riemann came to discovering the theory of gravity 60 years before Einstein."[21] In one way or other, speculations about the physics of Flatland have had profound consequences for us all.

Einstein's "discovery" of a fourth dimension must surely rate as one of modern science's most amazing findings. With this discovery, man was now in a position (like the Square in Abbott's tale) to see his world from a new perspective. But as the Square said to Lord Sphere, why stop at *four* dimensions? With our vision thus expanded, might we too not "resolve that our ambition [should] soar" onward and upward to higher dimensions still? And since human beings are as naturally curious as Squares, indeed it was not long before someone began to dream about a *fifth* dimension. In the 1920s, a young Polish mathematician had the bright idea that if the force of gravity could be explained by the geometry of four-dimensional space, then perhaps he might be able to explain the electromagnetic force by the geometry of five-dimensional space. With this seeming science fiction fantasy begins one of the most curious episodes in the history of space.

If Riemann was a maverick in the history of science, Theodr Kaluza was decidedly an oddity. An obscure mathematician at the University of Königsberg (in what is now Kaliningrad in the former Soviet Union), Kaluza was convinced that Einstein's approach to gravity could be expanded and enhanced. In particular, he wanted to apply Einstein's approach to the electromagnetic force—the force responsible for electricity, magnetism, and light. Along with Riemann, in fact, Kaluza believed that electromagnetism must also be the result of curvature (or ripples) in a higher-dimensional space. But the problem Kaluza faced was that there did not seem to be any more dimensions left. With three of space and one of time, nature's stock seemed to be exhausted.

Yet Kaluza was not a man to be deterred by such prosaic objections. In an audacious move he simply rewrote Einstein's equations of general relativity in five dimensions. Lo and behold, when he did so it turned out that these five-dimensional equations contained within them the regular four-dimensional equations of relativity, plus an extra bit which turned out to be precisely the equations of electromagnetism. In effect, Kaluza's five-dimensional theory consisted of two separate pieces that fitted together like a jigsaw puzzle—Einstein's theory of gravity and

Maxwell's theory of electromagnetism (the field equations of light).

Another way of understanding this "mathematical miracle," says physicist Paul Davies, is that "Kaluza showed that electromagnetism is actually a form of gravity." Not the regular gravity of everyday physics, but "the gravity of an unseen [fifth] dimension of space."[22] In 1919 Kaluza sent a paper on all this to Einstein. So stunned was the great physicist by the young Pole's radical addition of an extra dimension that like Lord Sphere in Abbott's Flatland, he was appalled. For two years Einstein apparently refused to answer Kaluza's letter. But the whole construction was so mathematically elegant he could not get it out of his mind, and finally in 1921 he became convinced of the importance of Kaluza's ideas and submitted the paper to a scientific journal.

Ironically, it was the very beauty of Kaluza's construction that so shook Einstein, and many other physicists. Was this five-dimensional space of Kaluza's "just a parlor trick? Or numerology? Or black magic?"[23] It was all very well to propose that *time* was a fourth dimension (for that, after all, is a real aspect of our physical experience), but what on earth was this supposed *fifth* dimension? If Kaluza's equations were to be taken seriously—and not just as mathematical chicanery—then the awkward question arose: Where is this extra dimension? Why don't we see it?

To this query Kaluza had a disarmingly simple answer. He declared that the extra dimension is so small it escapes our normal attention. The reason we don't see, he said, it is because it is microscopic. To understand this proposal, again it is helpful to resort to a lower-dimensional analogy. Imagine this time that you live on a line, what we might call Lineland, the one-dimensional sibling of Flatland. As a dot in this linear universe, you can travel up and down your line, always remaining in a single dimension. Now suppose that one day a scientist in your Lineland announces she has discovered an extra dimension and that your universe is really *two*-dimensional. At first you think she is mad. Where is this other dimension? you ask. Why can't we see it? But then the scientist explains that in fact you don't live on a line, but on a very thin *hose*. Each point of your line universe is not really a point, but a tiny *circle*, one so small that you never noticed it. Taking this extra microscopic dimension into account, your world is not a line, but really a two-dimensional cylindrical *surface*.

This was the essence of Kaluza's explanation for his fifth dimension. According to him, every point in our three dimensions of space is actually a tiny circle, so that in reality there are *four* dimensions of space, plus one of time, making a total of five. In 1926 the Swedish physicist Oskar Klein made improvements to Kaluza's theory which enabled him to calculate the size of this tiny hidden dimension. According to Klein's calculations, it was no wonder we had not observed the extra direction because it is absolutely minute. Its circumference was just 10^{-32} centimeters—a hundred billion billion (10^{20}) times smaller than the nucleus of an atom!

So small was Kaluza's dimension that even if we ourselves were the size of atoms we would *still* not notice it. Yet this tiny dimension could be responsible for all electromagnetic radiation: light, radio waves, X rays, microwaves, infrared, and ultraviolet. A powerful punch indeed for something so small. Unfortunately, the Kaluza-Klein dimension was so small there was no way of measuring it directly. Even our largest accelerators today still cannot measure things on such a minute scale. So what then are we to make of Kaluza's vision? Is this fifth dimension physically real? Or is it just an elegant mathematical fiction?

Kaluza himself insisted that the beauty of his theory could not "amount to the mere alluring play of a capricious accident."[24] He firmly believed in the reality of his fourth spatial dimension. He knew his tiny dimension could not be tested directly, so he decided instead to conduct an experiment of his own to test the general correspondence between theory and reality. The test case he chose was not anything from the realm of physics, but the art of swimming. As someone who could not swim, Kaluza decided he would learn all he could about the *theory* of swimming and when he had mastered that then he would test this theoretical framework against the reality of the sea. Giving himself over to the project, he diligently studied all aspects of the aquatic art until finally he felt he was ready. Now, trunks in hand, the young Pole escorted his family to the seaside for the crucial test. With no prior experience, in front of the assembled Kaluza clan, Theodr hurled himself into the waves . . . and lo and behold he could swim! Theory had been born out by practice in the real world. Could the tiny dimension *also* be there in the real world?

Unfortunately, if in Kaluza's own mind the swimming experiment supported a general correspondence between theory and reality, few others were willing to embrace the idea of an unseen and unmeasurable fifth dimension. Sadly, after an initial flurry of interest, the physics community turned away. Yet the startling elegance of Kaluza's equations raised an uneasy question: How many dimensions of space are there *really* in the world around us?

As happens so often in the history of science, it was not in fact a new question. As long ago as the second century, Ptolemy had considered the matter and had argued that no more than three dimensions are permitted in nature. Kant also had argued that three dimensions are inevitable. In this he could call upon the support a good deal of hard science. For instance it is well known that gravity and the electromagnetic force both obey "inverse square laws"—the strength of the force drops off according to the square of the distance. As early as 1747, "Kant recognized the deep connection between this law and the three-dimensionality of space."[25]

It turns out that in anything other than three dimensions, problems quickly arise with inverse square forces. For example, in four or more spatial dimensions, gravity would be so strong that planets would spiral into the sun; they would not be able to form stable orbits. Similarly, electrons would not be able to form stable orbits around nuclei.[26] Hence atoms could not form. It can also be shown that in four spatial dimensions, waves cannot propagate cleanly. From these physical facts, Kant and others had concluded that we *must* live in a universe with just three spatial dimensions.

But all these arguments had assumed that any extra dimensions would be fully extended like the regular three. If an additional dimension was tiny, however, it would *not* affect the regular functioning of gravity, electricity, and wave propagation. On the large scale, such a universe would operate as if there were just three dimensions; only on the *microscopic* scale would the extra one reveal itself. In other words, our universe could function properly with five dimensions.

If Kaluza was right, and such a thing *did* exist, it would pack a very potent punch. "Viewed this way, there [would be] no forces at all, only warped five-dimensional geometry, with particles meandering freely in a landscape of structured nothingness."[27] It was a very beautiful idea, but for over half a century most physicists paid no more attention to Kaluza than to Hinton or Ouspensky, and the fifth dimension seemed little less than an oddity of mathematical mysticism. Then suddenly in the 1980s that began to change when new developments in particle physics began to suggest that Kaluza might just have been onto something.

By the 1980s, two new forces of nature had been discovered. In addition to gravity and electromagnetism, there was now the *weak nuclear force* and the *strong nuclear force*. These forces are what holds atomic nuclei together, hence they are responsible for keeping matter stable. With these nuclear powers, the basic "forces of nature" had expanded in number from two to four. Today physicists feel confident that this set—gravity, electromagnetism, the weak force, and the strong force—represent the full complement of our physical universe. But what really began to excite them was the idea that all four might be just different aspects of a single overarching force—a kind of unifying *super-force*.

The idea of an underlying unity among all four forces of nature was so thrilling to many theoretical and particle physicists they were prepared to try anything to realize this vision. Many attempts were made to find a unifying theory, but after a decade of failure, they began to realize that desperate measures might be called for. At this point they began to look again at Kaluza. After all, he had been able to unify gravity and the electromagnetic force; perhaps his approach might be able to unify all four forces? Now, the idea of unseen hidden dimensions reared its head with a vengeance, for while Kaluza had been able to explain electromagnetism by adding just one more dimension to Einstein's equations, physicists found that in order to accommodate the weak and strong forces they had to add another *six* dimensions of space—bringing the total number of dimensions to *eleven!* As before, all these extra dimensions are microscopic—tiny little curled-up directions in space that can never be detected by human senses.

The picture that has emerged over the past decade is thus of an eleven-dimensional universe, with four extant, or large, dimensions (three of space and one of time), and seven microscopic space dimensions all rolled up into some tiny complex geometric form. On the scale that we humans experience, the world is four-dimensional, but underneath, say these new "hyperspace" physicists, the "true" reality is eleven-dimensional. (Or, according to some of the latest theories, maybe ten-dimensional.)

Perhaps the most radical feature of this eleven-dimensional vision is the fact that it explains not only all the forces, but *matter* also, as a by-product of the geometry of space. In these extended Kaluza-Klein theories, matter too becomes nothing but ripples in the fabric of hyperspace. Here, subatomic particles are also explained by the properties of the seven curled-up dimensions. One of the major projects of theoretical physics over the past two decades has been to articulate precisely how the curling up of these extra spatial dimensions occurs. Unfortunately there are an enormous number of possible topologies for a seven-dimensional space, and so far it has proved impossible to tease out which ones (if any) correspond to the real world we live in. Part of the problem, again, is that all these dimensions are too tiny to be measured directly, so any such theories can only be tested indirectly—if at all. Nonetheless, hyperspace physicists are confident that they will find the correct one.

We have looked at how the curvature of space can produce the effect of physical forces such as gravity; let us consider now the even more radical idea that the curvature of space may also be responsible for *matter*. Forces such as gravity and magnetism (which travel through thin air) have always, in a sense, been closely allied with space, but how could matter—the concrete stuff of our flesh and bones—arise from the non-substance of space?

At first glance the whole notion seems absurd, but once again the idea of matter as ripples in space is actually quite old. As early as the 1870s Riemann's English disciple William Clifford delivered an address to the Cambridge Philosophical Society "On the Space Theory of Matter."[28] Taking Riemann's ideas further even than the master himself, Clifford put forward the view that particles of matter were just tiny kinks in the "fabric" of space. A more sophisticated version of the same idea arose early in our own century when physicists began to think about wormholes. Original interest in wormholes was not in the large-scale ones that would so excite science fiction writers, but in microscopic wormholes that might be associated with subatomic particles. A host of physics luminaries from Einstein to Hermann Weyl "wondered whether all fundamental particles might not actually *be* microscopic wormholes."[29] In other words, just "the products of warped spacetime."

Einstein in particular became obsessed with the notion that matter might be ripples in space, and he spent the last thirty years of his life trying to extend the equations of general relativity in this direction. He called this dream a "unified field theory" and his failure to find such a theory was the greatest disappointment of his life. According to Kaku, "to Einstein the curvature of spacetime was like the epitome of Greek architecture, beautiful and serene."[30] But he regarded matter as messy and ugly. He likened space to "marble" and matter to "wood," and he desperately wanted a theory that could transform ugly "wood" into beautiful "marble."

Neither Clifford nor Einstein had the mathematical tools to achieve the difficult synthesis of matter and space—above all they were trying to work with just four dimensions. Today physicists know that if matter *is* to be incorporated into the structure of space, it must be achieved with a higher-dimensional theory. In such a theory, matter, like force, would not be an independent entity, but a secondary by-product of the totalizing substrate of space. Here, everything that exists would be enfolded into the bosom of hyperspace. Theories that attempt to do this are sometimes known by the modest nickname "theories of everything," commonly referred to as TOEs. In a successful TOE, every particle that exists would be described as a vibration in the microscopic manifold of the extra hidden dimensions. Objects would not be *in space*, they would *be space*. Protons, petunias, and people—we would all become patterns in a multidimensional hyperspace we cannot even see. According to this conception of reality, our very existence as material beings would be an illusion, for in the final analysis there would be nothing but "structured nothingness."

With a hyperspatial "theory of everything" we thus reach the apogee of a movement that began in the late Middle Ages: The elevation of space as an ontological category is now complete. As we have seen, in the Aristotelian world picture, space was a very minor and unimportant category of reality—so unimportant that Aristotle didn't really have a theory of "space" per se but strictly speaking only a theory of "place." With the emergence of Newtonian physics in the seventeenth century, the status of space was raised so that along with matter and force it became one of *three* major categories of reality. Now, at the close of the twentieth century, space is becoming the *only* primary category of the scientific world picture. Matter and force, which in Newtonian physics were really above space in ontological status, have now been relegated to secondary status, with space alone occupying the primary rung of the real. It is a little-remarked-upon feature of modern Western physics that one way of characterizing the enterprise is by the gradual ascent of space in our existential scheme. The final triumph of this invisible, intangible entity to the *ultimate essence* of existence is surely one of the more curious features of any world picture.

Hyperspace physicists' intensely geometric vision of reality also marks the final chapter of the saga begun by Giotto and the geometer-painters of the Renaissance. Here in TOE physicists' equations would be the ultimate "perspective" picture of the world, a vision in which *everything* is refracted through the clarifying prism of geometry. If, as Plato famously declared, "God ever geometrizes," here would be the last word on divine action. As the apotheosis of Roger Bacon's "geometric figuring," a hyperspatial "theory of everything" would be, quite simply, a twenty-first-century realization of a thirteenth-century dream.

In another way also a "theory of everything" would be the ultimate perspective picture of our universe, for this picture too has a *single point* from which the whole world-image *originates*. Physicists call it the big bang. According to hyperspace physics, at the initial split second of creation the entire universe was condensed into a microscopic point containing all matter, force, energy, and space. At this quintessential point, however, matter, force, energy, and space were not yet separated from one another, but were united in a single hyperspace substrate. In other words, at the split second of creation *everything* was folded within the all-embracing oneness of "pure" eleven-dimensional space. From this point of hyperspatial unity, the universe then unfolded.

As the single point from which the physicists' world picture *originates*, the big bang is a scientific equivalent of the perspective painters' "center of projection." It is the point at which all "lines" in the hyperspace universe converge. This is the place, then, where TOE physicists would dearly like to "stand." Just as the viewer of a perspective painting gets the most dramatic effect when standing in the place from which the artist constructed the image, so a hyperspace physicist could see his world picture most clearly if he "stood" at the cosmic center of projection—the big bang.

It is in search of this particular "point of view" that physicists build ever larger particle accelerators. The higher the energy one can generate in an accelerator, the closer one gets to "melting" together the four separate forces, and thus the more one can see of the underlying hyperspatial unity. In a very real sense, particle accelerators are tools for exploring higher-dimensional space, and the final goal with such machines is to glimpse once more the initial point of "pure" eleven-dimensional hyperspace. Physicists speak about this initial period of hyperspace unity as the time when there was "perfect symmetry" between all eleven dimensions. What they want to do is to glimpse for themselves this original perfect symmetry. Ironically, while artists long ago abandoned Renaissance aesthetics, those classical ideals of beauty live on in physicists' dream of a "theory of everything." Like the Renaissance painters, TOE physicists also hold mathematical symmetry as the highest aesthetic ideal. It is their dream, their goal, and, it has even been said, their "Holy Grail."

Given the pedigree of hyperspace physicists' world picture, we should not be surprised to find that it results in the same kind of homogenizing tendencies as those witnessed in Renaissance art. As psychologist and historian Robert Romanyshyn has noted, in the space of linear perspective "all things, regardless of what they are and regardless of the context to which they belong, are equal and the same."[31] Indeed, as we saw in Chapter Two, that equality was precisely the point of perspectival representation. If this is true of perspectival imagery, how much more so of hyperspace physicists' world picture, in which all things are literally *the same* — everything being just a manifestation of hyperspace. Here, everything that is "is reduced to the same plane or level of reality."[32] Just like perspective painting, the hyperspatial "theory of everything" gives us "a vision which perceives everything belonging to the same plane," indeed to the selfsame existential category. No longer do we have even a distinction between matter and space, for now there is just one, and only one, category of existence. Seven hundred years after Giotto, the geometric *leveling* of the world, prefigured in his Arena Chapel Christ cycle, has thereby reached its conclusion. All gradients of reality, all existential distinctions, have finally been annihilated. Homogenization has won the day.

As we saw in Chapters Two and Three, the original geometrizing of space from the fourteenth through seventeenth centuries created a world picture in which the physical realm came to be seen as the totality of reality. In this original physicalist vision there was no longer any place for a realm of soul or spirit because physical space was extended to infinity. How much less a place, then, in the new hyperspace vision where physical space is not only infinite, in itself it has become the totality of the real. By positing everything as empty space curled into patterns, hyperspace physics profoundly denies any "other" levels of reality. With this vision, "any sense of the world as a reality of multiple levels simultaneously co-existing [becomes] the stuff of fancy and of dream."[33]

But the problem is more profound even than the denial of other planes of reality, because by making space the *only* category of the real we also deny what Edwin Burtt terms "time as something lived."[34] With a "theory of everything" time is effectively frozen, for here it becomes just another dimension of space. Subdued by the cool hand of geometry, "lived time," "flowing time," "variable time" are all annihilated in the crystalline grip of eleven-dimensional symmetry.[35] In hyperspace physicists' world picture, time is no longer an attribute of *subjective human experience*; it becomes just an artifact of mathematical manipulation. Thus not only are the atoms of our bodies stripped of independent status and reduced to spatial origami, our most fundamental experience of time as something lived and personal is annihilated. In the eleven-dimensional manifold of various "theories of everything," our very being disappears into "structured nothingness." We are dissolved in space.

With just one ontological category of reality, there can only ever be one plane of reality—and in TOE physicists' vision that is the physical plane. Whatever hopes Ouspensky et al. may have had for higher-dimensional space as a spiritual haven, contemporary scientific accounts of hyperspace remain purely physical. In short, with a hyperspatial "theory of everything," our world picture is reduced fully and finally to a seamless *monism*. The movement that we have been tracing in this book from the medieval *dualistic* vision of physical space and spiritual space has thus reached its climax. Here, everything is equal, everything is homogeneous, everything is space. Paradoxically, this new monism privileges neither body nor spirit, for with matter itself being just a by-product of space, body too is ultimately annulled. What remains is just empty space curling in on itself.

Interestingly, but perhaps not surprisingly, some physicists have attempted to interpret the "theory of everything" itself in a spiritual sense. It is this theory that Stephen Hawking has so famously associated with the "mind of God." And Hawking is by no means alone in equating a TOE with God; other physicists also have been doing this, including most notably Paul Davies and

Nobel prize–winning particle physicist Leon Lederman. Just what is the "God" of TOE physics is far from clear, but this deity seems to be little more than a set of equations. Yet along with many non-scientists, these physicists have sensed that for many people today a purely physicalist world picture is *not* satisfying. By attempting to equate their hyperspatial vision with a "God," these men are trying to reinfuse both the scientific world picture and space itself with some kind of spirituality.

This may be an admirable goal, but it is one that I suggest is doomed to failure. What we have here is a similar move to that which we saw at the end of Chapter Three with Newton and More's attempts to infuse a sense of religiosity into the mechanistic world picture by divinizing Euclidian space. Just as that effort ultimately failed, so too, I suggest, will any attempt to divinize hyperspace. As we remarked earlier, physical space *cannot* be the foundation for a genuine theology—Christian or otherwise. For those who wish to see reality as more than a purely physical phenomena—and I include myself in this camp—the way forward is not to try to divinize physicists' latest conceptions of space, but rather to understand *their* picture as just *one part* of the whole.

Before we go on, I want to stress here that as a student of physics I am deeply fascinated by the hyperspace saga. As an aesthetic exercise especially I think it is extraordinary. I am not questioning the validity of their mathematical vision per se, which like Renaissance perspective art interests me greatly. What I *do* want to question is the notion that this hyperspace vision constitutes the *totality* of reality. What I challenge then is not the science, but rather the totalizing interpretation of what this science *means*.

Let us assume for a moment that TOE physicists had succeeded in their program, and that they had on their blackboards a set of equations describing a multidimensional space that encoded all the known particles and forces. That such a theory would be an extremely beautiful picture I have every confidence. But beauty is not necessarily truth, or at least, not the *whole truth*. What would such a theory *mean*? How should it be *interpreted*?

The answer, I suggest, is essentially the same as to the question of how we should interpret a Raphael painting. While it is true that the *Disputa* (for example) is in many ways a "realistic" image, nobody today would accept it as a depiction of the real world. Even the lower part of the image that deals strictly with the physical realm is still highly stylized and in many ways unnatural. For all Raphael's perspective brilliance, paintings of the quatrocento have a rather stilted feel. There is something too geometrically perfect about these images. The world of our *experience* does not possess that pristine clarity; and even when disguised by the softening tones of chiaroscuro, High Renaissance images possess a geometrical rigidity that is wholly unnatural. Glorious though a Raphael or a Leonardo painting may be, you know and I know that reality is *not* like this—at least not totally.

Certainly such images do encode *something* of reality. We do immediately recognize the scenes portrayed, and we do feel drawn into the virtual spaces beyond the picture frame precisely because there is a convincing illusion of physical space here. We do recognize the faces, who do look like real people. Who can say they do not feel that Mona Lisa stares back at them? Who can say they are not moved by a Piero della Francesca portrait or a Raphael madonna because they do seem so human? Yet in the end we are not fooled by perspectival imagery. We recognize that while it captures some aspects of reality, it does not capture *all aspects*.

That is precisely what Raphael himself acknowledged in the *Disputa* when he abandoned perspective at the top of his image. As he reminds us, the totality of reality cannot be suborned under the banner of geometry. Raphael, of course, was referring to the particular reality of Christian soul-space, but even if we don't think in Christian terms, reality is *not* totally reducible to the laws of physics. Love, hate, fear, jealousy, delight, and rage—none of these can be accounted for by hyperspace equations. In a very profound sense Descartes' *res cogitans* is still the skeleton in the closet of modern science. Try as the new physicalists will to convince us that the epistemic battles are over, it is evident that they are not. Whether we see ourselves in religious or secular terms, we humans are creatures of *psyche* as well as soma, and no hyperspace theory will help to illuminate *that*.

Just as perspectival representation captures some aspects of reality, so also a "theory of everything" would describe many real phenomena. Objects *do* fall to the ground as Newton's law of gravity suggests; light *does* bend around the sun as general relativity predicts; electric and magnetic fields *do* interact in the manner implied by Maxwell's equations; beta radioactive decay *does* occur as the weak theory predicts; and inside accelerators, protons *do* behave as the theory of the strong force suggests. To find a single theory that unified all these forces would be an extraordinary achievement; it would give us a truly astonishing "perspective" on our world. But like a quatrocento painting, such a depiction would not describe the totality of reality—and it is utterly naive to suggest that it will.

Just as artists have long recognized the limited scope of linear perspective, so too we must recognize the limited scope of physicists' "pictures" of the world. As with perspective, modern physics filters the world through a rarefied lens—specifically, it filters the world through a mathematical lens. What emerges from this highly selective process are deeply *constructed* images of the world, very much like linear perspective images. These images are "real" to the extent that physicists' equations allow us to make extremely accurate predictions about all sorts of phenomena, from the way gravity behaves to the action of subatomic particles. There *is* a sense in which, to use physicist Eugene Wigner's famous phrase, mathematics is "unreasonably effective." Just as mathematics proved "unreasonably effective" in certain aspects of artistic practice, so too it is incredibly effective in describing many physical phenomena. But just because the language of mathematics can be used to accurately describe many aspects of the world does not mean it can provide us with a *total* world picture. We must recognize that this is true in science as well as in art.

Thus we return to our point of departure: to the man who set us on this path in the first place, Giotto di Bondone. Although, as we saw in Chapter Two, Giotto did as much as anyone to develop perspectival vision, he never mistook his "geometric figurings" for the totality of the real. Alongside his seminal explorations of "realistic" figuration, Giotto also painted the profoundly medieval *Last Judgment,* which eschewed entirely the new geometric mode of representation. Here the great painter insisted on a multileveled vision of reality. Living on the cusp of two eras, Giotto could appreciate the power of geometric figuring, while at the same time never losing sight of other planes of the real. The trouble with some hyperspace physicists today is that they want to assert *their* geometric figurings as the *totality* of the real. They want to insist on the all-embracing power of their vision alone.

This is nothing more or less than the latest version of the old physicalist hegemony whose rise we have been charting throughout this book. But history has dealt these new physicalists a wild card, for just when it seems that their hyperspace vision is on the verge of completion, a *new* space is beginning to emerge—one that stands quite outside their equations. As the twentieth century draws to a close, completing movements set in train by the aesthetic revolution of the fourteenth and fifteenth centuries and by the scientific revolution of the sixteenth and seventeenth, the modern mathematical triumph over physical space is being overtaken by a new, and most unexpected, revolution. Beyond the bounds of hyperspace—unreachable by *any* number of extra dimensions— the digital universe of the Internet explodes into being with the irrepressible force of its own "big bang." As we approach a new millennium the new spatial frontier is not hyperspace, but *cyberspace.*

CYBERSPACE

With the exponential force of its own big bang, cyberspace is exploding into being before our very eyes. Just as cosmologists tell us that the physical space of our universe burst into being out of nothing some fifteen billion years ago, so also the ontology of cyberspace is ex nihilo. We are witnessing here the birth of a new domain, a new space that simply did not exist before. The interconnected "space" of the global computer network is not expanding into any previously existing domain; we have here a digital version of Hubble's cosmic expansion, a process of space creation.

Like physical space, this new "cyber" space is growing at an extraordinary rate, increasing its "volume" in an ever-widening "sphere" of expansion. Each day thousands of new nodes or "sites" are added to the Internet and other affiliated networks, and with each new node the total domain of cyberspace grows larger. What increases here is not volume in any strictly geometrical sense—yet it is a *kind* of volume. In cyberspace each site is connected to dozens, or even thousands, of others through software-defined "hot buttons." These digital connections link sites together in a labyrinthian web that branches out in many "directions" at once. In describing cyberspace we might use the words "web" and "net," which classically are two-dimensional phenomena, but even the most neophyte surfer knows that cyberspace cannot be constrained by two axes. This new, enigmatic, space is the subject of our remaining three chapters.

Cyberspace is not just expanding, it is doing so exponentially. In this sense also its genesis parallels that of physical space. According to the latest theories of cosmology, before the smoothly expanding universe we see today there was an early phase of wildly excessive expansion that physicists refer to as the "inflationary" period. During this phase, space swelled from a microscopic point smaller than a proton to the size of a grapefruit in a fraction of a second. In this larval stage, the rudiments of large-scale cosmic structure were laid down, the body-plan, as it were, for the galactic web that constitutes our universe today.

Right now cyberspace is going through its own inflationary period. In the past fifteen years, the Internet has swelled from fewer than a thousand host computers to more than thirty-seven million—and growing by the day. Because each new node becomes in itself a hub from which further nodes might sprout, the greater the number of nodes the greater the possibility for even more expansion. In this seminal inflationary phase the large-scale structure of the cyber-domain is also being formed.

The exponential pattern of cyberspatial growth is evidenced by even a most cursory history. The dawn of cyber-creation—the first quantum flicker, as it were, of a new domain tunneling into being—can be traced to California in 1969. That year saw the formation of the world's first long-distance computer network, the ARPANET, funded by the U.S. Department of Defense (DOD) through its Advanced Research Projects Agency (ARPA). In October 1969, technicians from the Boston-based firm Bolt Beranek and Newman linked together, via specially laid telephone lines, two computers hundreds of miles apart, one at UCLA, the other at the Stanford Research Institute. By the end of the year two more nodes had been added to this nascent net—the University of California at Santa Barbara and the University of Utah—making a network of four sites.[1]

By the next year, write computer historians Katie Hafner and Matthew Lyon, "the ARPA network was growing at a rate of about one node per month,"[2] and by August 1972 it contained twenty-nine nodes located in universities and research centers across the USA.[3] In these early years, when maintaining a site cost more than $100,000 per annum (with all the money coming from the DOD), growth was necessarily incremental.[4] Indeed, by 1979, a decade after the first two sites were connected, there were still just sixty-one ARPANET sites.

The advantages of what was already being called "the Net" were, however, becoming evident, and more and more people—especially computer scientists—were calling for online access. But as a research project of the Defense Department, the ARPANET was not easily available to anyone outside ARPA's direct circle. Clearly there was need for a civilian network as well. To that end, in 1980 the National Science Foundation decided to sponsor a network to connect the growing number of computer science departments around the country—the CSNET. Though separate, the two networks were interconnected so that members of each could communicate with one another. During the eighties, other networks also were connected to the ARPANET, creating a global network of networks. The growing desire to communicate *between* networks brought about the need for a standardized set of procedures that would enable all networks to pass information amongst themselves—what came to be called an "Internet Protocol." From this originally technical term the "Internet" would get its name.

Still the Net remained a rarefied domain. In the early eighties few people outside the military and the academic field of computer science had any network access, and few Americans were even aware that "cyberspace" existed.[5] The word itself was only coined in 1984, in William Gibson's seminal cyberpunk novel, *Neuromancer*. In 1985, however, the expansion of cyberspace shifted into a higher gear. Following the success of CSNET, the National Science Foundation made the further decision to build a national "backbone" network to serve as the foundation for a series of regional networks linking universities around the country. Replacing the outdated ARPANET, this NSFNET was the basis of what soon became the Internet.

The creation of the NSFNET marks a turning point in the history of cyberspace: Here was the start of cyberspatial inflation. Since then the pace of growth has accelerated rapidly, outstripping the wildest imaginings of its creators. By late 1998, as I write, the World Wide Web (which is the most public component of the Internet) has over 300 million pages. So much volume is being added to the World Wide Web that major cataloging services such as Yahoo and AltaVista estimate their libraries have logged only 10 percent of the total. Inflationary growth on the Web is now so extreme that experts worry they will never be able to keep track of it all.

A hitherto nonexistent space, each year this new digital domain plays a greater role in more and more people's lives. Like many "netizens," I now have e-mail correspondents around the world. People with whom it would be difficult to communicate in the flesh are often readily available online, especially if they work in the academic arena. Almost all academic institutions, research centers, and major libraries in the United States now have Web sites. Through my computer I can access the catalog of the Library of Congress and that of UCLA, which is physically located just a mile from where I live. In the not-too-distant future, the texts themselves will also be online; as already is the content of many magazines and newspapers. Why buy the *New York Times* on paper when you can read it online for free? Moreover, in the new publishing paradigm now emerging, many publishers eschew hard copy entirely and only publish online.

Businesses too are staking out a presence in cyberspace. Seemingly every corporation from IBM and Nike on down now sports a Web site packed with corporate PR and product information. Included in an increasing number of sites is also the ability to purchase online. Clothes, books, cosmetics, airline tickets, and computer equipment (to name just a few items) can now be bought over the Net. According to a recent Commerce Department report, ten million people in the United States and Canada had bought something online by the end of 1997. The report estimates that electronic commerce should reach $300 billion by 2002. The virtual mall has arrived.

Whatever the vision of the Internet's founders, cyberspace has long since burst the husk of its academic seedpod. These days every second college kid in America has his or her own home page, spawning what must be the largest archive ever of the adolescent mind. A growing number of families are also "moving" into cyberspace, keeping loved ones posted online with digitized snapshots of their summer holidays. With the advent of automated Web site–authoring software, the family home page is destined to become as ubiquitous as the old photo album—and a lot more public.

Most prominently, cyberspace is a new place to socialize and play. Chat rooms, newsgroups, IRC channels, online conferences and forums, and the fantasy worlds known as MUDs—all seem to promise almost infinite scope for social interaction. Moreover, in cyberspace one can readily search for friends with similar interests. As online pioneer Howard Rheingold has written, while "you can't simply pick up a phone and ask to be connected with someone who wants to talk about Islamic art or California wine, or someone with a three-year-old daughter or a forty-year-old Hudson, you can, however, join a computer conference on any of these topics."[6] The level of discussion in many public forums may well be highly variable, but serious *private* online discussion groups abound on a vast array of topics, from biblical exegis to particle physics, from *The Divine Comedy* to the big bang.

As of mid 1998, there are one hundred million people accessing the Internet on a regular basis and it is estimated that in the next decade there will be close to a billion people online. With three hundred million pages already on the World Wide Web, it is currently growing by a million pages a day. In just over a quarter century, this space has sprung into being from nothing, making it surely the fastest-growing "territory" in history.

In a very profound sense, this new digital space is "beyond" the space that physics describes, for the cyber-realm is not made up of physical particles and forces, but of *bits* and *bytes*. These packets of data are the ontological foundation of cyberspace, the seeds from which the global phenomena "emerges." It may be an obvious statement to say that cyberspace is not made up of physical particles and forces, but it is also a revolutionary one. Because cyberspace is not ontologically rooted in these physical phenomena, it is *not subject to the laws of physics*, and hence it is not bound by the limitations of those laws. In particular, this new space is not contained within physicists' hyperspace complex. No matter how many dimensions hyperspace physicists add into their equations, cyberspace will remain "outside" them all. With cyberspace, we have discovered a "place" *beyond hyperspace*.

We should not underestimate the importance of this development. The electronic gates of the silicon chip have become, in a sense, a metaphysical gateway, for our modems transport us out of the reach of physicists' equations into an entirely "other" realm. When I "go" into cyberspace I leave behind both Newton's and Einstein's laws. Here, neither mechanistic, or relativistic, or quantum laws apply. Traveling from Web site to Web site, my "motion" cannot be described by *any* dynamical equations. The arena in which I find myself online cannot be quantified by *any* physical metric; my journeys there cannot be measured by *any* physical ruler. The very concept of "space" takes on here a new, and as yet little understood, meaning, but one that is definitively beyond physicists' ken.

Ironically, cyberspace is a technological by-product of physics. The silicon chips, the optic fibers, the liquid crystal display screens, the telecommunications satellites, even the electricity that powers the Internet are all by-products of this most mathematical science. Yet if cyberspace could not exist without physics, neither is it bound within the purely physicalist conception of the real. In the parlance of complexity theory, cyberspace is an *emergent phenomena,* something that is more than the sum of its parts. This new "global" phenomena *emerges* from the interaction of its myriad interconnected components, and is not reducible to the purely physical laws that govern the chips and fibers from which it indubitably springs.

All this may sound rather radical, and many cyberspace enthusiasts have suggested that nothing like cyberspace has existed before. But on the contrary there is an important historical parallel here with the spatial dualism of the Middle Ages. As we have seen, in that time Christians believed in a physical space described by science (what they called "natural philosophy") and a nonphysical space that existed "outside" the material domain. This nonphysical space metaphorically *paralleled* the material world, but it was not contained within physical space. Although there were connections and resonances between the two spaces, medieval spiritual space was a separate and unique part of reality from physical space.

So too the advent of cyberspace returns us to a *dualistic* theater of reality. Once again we find ourselves with a material realm described by science, and an immaterial realm that operates as a different plane of the real. As with the medieval world picture, there are connections and resonances between these two spaces. Commentator N. Katherine Hayles has noted, for example, that one cannot experience cyberspace at all except through the physical senses of the body: the eyes that look at the computer screen or at the stereoscopic projections of virtual reality headsets, the hands that type the commands at the keyboard and control the joysticks, the ears that hear the Real Audio sound files. Yet while physical space and cyberspace are not entirely separate, neither is the latter *contained* within the former.

In some profound way, cyberspace is *another* place. Unleashed into the Internet, my "location" can no longer be fixed purely in physical space. Just "where" I am when I enter cyberspace is a question yet to be answered, but clearly my position cannot be pinned down to a mathematical location in Euclidian or relativistic space—not with any number of hyperspace extensions! As with the medievals, we in the technologically charged West on the eve of the twenty-first century increasingly contend with a two-phase reality.

But what does it mean to talk about this digital domain as a "space" at all? What kind of space is it? Some might object that the online arena is just a vast library—or less generously, a vast soup—of disconnected information and junk. And certainly there is a lot of junk online. Nonetheless, it is important to recognize the genuinely spatial nature of this domain. Whatever its *content* may be, a new *context* is coming into being here; a new "space" is evolving.

What is at issue, of course, is the meaning of the word "space" and what constitutes a legitimate instance of this phenomena. I contend that cyberspace is not only a legitimate instantiation of this phenomena but also a socially important one. In the "age of science" many of us have become so habituated to the idea of space as a purely physical thing that some may find it hard to accept cyberspace as a genuine "space." Yet Gibson's neologism is apposite, for it captures an essential truth about this new domain. When I "go into" cyberspace, my body remains at rest in my chair, but "I"—or at least some aspect of myself—am teleported into another arena which, while I am there, I am deeply aware has its own logic and geography. To be sure, this is a different sort of geography from anything I experience in the physical world, but one that is no less real for not being material. Let me stress this point: *Just because something is not material does not mean it is unreal*, as the oft-cited distinction between "cyberspace" and "real space" implies. Despite its lack of physicality, cyberspace is a real place. *I am there*—whatever this statement may ultimately turn out to mean.

Even in our profoundly physicalist age, we invoke the word "space" to describe far more than just the physical world. We talk about "personal space," and about having "room to move" in our relationships, as if there was some kind of relationship space. We use the terms "head space" and "mental space," and Lacanian psychoanalysts (following Freud) believe the mind itself has a spatial structure. Literary theorists discuss literary space and artists discuss pictorial space.

Contemporary scientists, for their part, now envisage a whole *range* of nonphysical spaces. Chemists designing new drugs talk about molecular space; biologists talk about evolutionary spaces of potential organisms; mathematicians study topological spaces, algebraic spaces, and metric spaces; chaos theorists studying phenomena such as the weather and insect plagues look at phase spaces, as indeed do physicists studying the motion of galaxies and the quantum behavior of atoms; and in a recent *Scientific American* article an epidemiological analysis of the spread of infectious diseases posited the idea of viral spaces. "Space" is a concept that has indeed come to have enormous application and resonance in the contemporary world.

Most obviously, the online domain is a *data space*. This was the concept at the core of Gibson's original cyberpunk vision. In *Neuromancer* and its sequels, Gibson imagined that when his "console cowboys" donned their cyberspace helmets, they were projected by the power of computer-generated three-dimensional illusionism into a virtual data landscape. Here, the data resources of global corporations were represented as architectural structures. The data bank of the Mitsubishi Bank, for example, was a set of green cubes, that of the "Fission Authority" was a scarlet pyramid. As a nice example of life imitating art, Tim Berners-Lee, the inventor of the World Wide Web, has said that his goal when designing the Web was to implement a global data space that could be accessed and shared by researchers around the world. We are yet to realize the full VR splendor of Gibson's original vision, but the essential concept of a global data space is already manifest in the World Wide Web.

But cyberspace has become much more than just a data space, because as we have noted much of what goes on there is *not* information-oriented. As many commentators have stressed, the primary use of cyberspace is not for information-gathering but for social interaction and communication—and increasingly also for interactive entertainment, including the creation of a burgeoning number of online fantasy worlds in which people take on elaborate alter egos.

What I want to explore in this first cyberspace chapter are the ways in which this new digital domain functions as a space for complex mental experiences and games. In this sense, we may see cyberspace as a kind of electronic *res cogitans*, a new space for the playing out of some of those immaterial aspects of humanity that have been denied a home in the purely physicalist world picture. In short, there is a sense in which cyberspace has become a new realm for the mind. In particular it has become a new realm for the imagination; and even, as many cyber-enthusiasts now claim, a new realm for the "self." To quote MIT sociologist of cyberspace Sherry Turkle: "The Internet has become a significant social laboratory for experimenting with the constructions and reconstructions of self that characterize postmodern life."[7] Just what it means to say that cyberspace is an arena of "self" is something we must examine closely, but the claim itself commands our attention.

The fact that we are in the process of creating a new immaterial space of being is of profound psychosocial significance. As we have been documenting in this book, any conception of "other" spaces being "beyond" physical space has been made extremely problematic by the modern scientific vision of reality. That problematizing is one of the primary pathologies of the modern West. Freud's attempt, with his science of *psychoanalysis*, to reinstate mind or "psyche" back into the realm of scientific discourse remains one of the most important intellectual developments of the past century. Yet Freud's science was distinctly individualistic. Each person who enters psychoanalysis (or any other form of psychotherapy), must work on his or her psyche individually. Therapy is a quintessentially lonely experience. In addition to this individualistic experience, many people also crave something communal—something that will link their minds to others. It is all well and good to work on one's own personal demons, but many people also seem to want a *collective mental arena*, a space they might share with other minds.

This widespread desire for some sort of collective mental arena is exhibited today in the burgeoning interest in psychic phenomena. In the United States psychic hot lines are flourishing, belief in an "astral plane" is widespread, and spirit chanelling is on the rise. In the latter case, the posited collective realm transcends the boundary of death, uniting the living and dead in a grand brotherhood of the ether. Meanwhile, *The X-Files* offers us weekly promises of other realities beyond the material plane, and bookstores are filled with testimonials describing trips to an ethereal realm of light and love that supposedly awaits us all after death. One of the great appeals of cyberspace is that it offers a *collective immaterial arena* not after death, but here and now on earth.

Nothing evinces cyberspace's potential as a collective psychic realm so much as the fantastic online worlds known as MUDs.[8] Standing for "multiuser domains" or originally "multiuser Dungeons and Dragons," MUDs are complex fantasy worlds originally based on the role-playing board game Dungeons and Dragons that swept through American colleges and high schools in the late seventies. As suggested by the "Dungeons and Dragons" moniker, the original MUDs were medieval fantasies where players battled dragons and picked their way through mazes of dungeons in search of treasure and magical powers. Today MUDs have morphed into a huge range of virtual worlds far beyond the medieval milieu. There is TrekMUSE, a Star Trek MUD where MUDers (as players are called) can rise through the ranks of a virtual Starfleet to captain their own starship. There is DuneMUD based on Frank Herbert's science fiction series, and ToonMUD, a realm of cartoon characters. The Elysium is a lair of vampires, and FurryMuck a virtual wonderland populated by talking animals and man-beast hybrids such as *squirriloids* and *wolfoids*.

Like good novels, successful MUDs evoke the sense of a rich and believable world. The difference is that while the reader of a novel encounters a world fully formed by the writer, MUDers are actively involved in an ongoing process of world-making. To name is to create, and in MUD worlds the simple act of naming and describing is all it takes to generate a new alter ego or "cyber-self." MUDers create their online characters, or personae, with a short textual description and a name. "Johnny Manhattan," for instance, is described as "tall and thin, pale as string cheese, wearing a neighborhood hat"; Dalgren is "an intelligent mushroom that babbles inanely whenever you approach"; and Gentila, a "sleek red squirriloid, with soft downy fur and long lush tresses cascading sensuously down her back." Within the ontology of these cyber-worlds, you *are* the character you create. As one avid player puts it, here "you are who you pretend to be."[9] Want to be a poetry-quoting turtle, a Klingon agent, or Donald Duck? In a MUD you can be.

MUDing is quintessentially a communal activity in which players become integrally woven into the fabric of a *virtual society*. Part of that process is the continuing evolution of the world itself. While the basic design of a MUD is determined by its program-mer creators, generally known as "wizards" or "gods," in most MUDs players can construct their own rooms or domiciles. Using simple programming commands, MUDers "build" in software or, simply with a textual description, a private space to their own taste. Personal MUD rooms span the gamut from a book-lined tree house, to a padded cell, to the inside of a television set. In some MUDs players can also build larger structures. Citizens of the Cyberion City space station in the MicroMUSE, for example, have built for themselves a science center, a museum, a university, a planetarium, and a rain forest.

Above all, a MUD is sustained by the *characters* who popu-late it. To use William Gibson's famous phrase, a MUD is a para-digmatic instance of the "consensual hallucination" of cyberspace.[10] Fantasy worlds (whether online or off) are always only as good as the imaginations holding them together, and in successful MUDs the other players are just as keen as you are to take your "squirriloid" nature seriously. As the Unicorn said to Alice on the other side of the looking glass: "If you'll believe in me, I'll believe in you." In successful MUDs everyone is striving for maximal conviction, both for their own character and for the world as a whole.

The interlocking imaginative and social mesh of a MUD means that actions taken by one player may affect the virtual lives of hundreds of others. As in the physical world, relationships build

up over time (not untypically over thousands of hours of online en-
gagement); trusts are established, bonds created, responsibilities
ensue. The very vitality and robustness of a MUD emerges from
the collective will of the group, wherein the individual cyber-self
becomes bound into a social matrix that is none the less real for
being virtual. When, as in some combat-based MUDs, a character is killed, often there is a strong sense of loss for the actual
human being who has spent hundreds of hours establishing the
character. "Gutted" is the word players use; because as Richard
Bartle, cocreator of the first MUD, explains, "it's about the only
one that describes how awful it is."[11]

What may at first may appear little more than juvenile fantasies—talking animals, space cadets, and Toon-town—can, however, turn out to be surprisingly complex domains of psychosocial
exploration. A MUDer friend of mine tells me that for her,
MUDing is a way to express sides of herself that she feels are not
sanctioned by the relentless "put on a happy face" optimism of
contemporary can-do America. MUDing allows out a darker, but,
she feels, a more "real" side of herself. For her MUDing is not so
much a game as a way to explore and express important aspects of
her "self," which (she feels) could not easily be exercised in flesh-
and-blood society. Turkle, who has been studying MUD cultures
since the early 1990s, notes that my friend's experience is not uncommon. As she writes, these fantasy environments may allow
"people the chance to express multiple and often unexplored aspects of the self."[12]

One parallel here is with masks. As actors and shamans attest,
masks are powerfully transformative objects. Hidden behind an er-
satz face, a man can "become" a wind devil, a monkey spirit, or an
ass. MUD descriptors are digital masks, fronts that may enable a
range of psychological expression and action, which many people
in modern societies may not have access to in their regular lives,
or which they do not feel comfortable unleashing in the flesh.

"Part of me," says one of Turkle's MUDers, "a very important part of me, only exists inside PernMUD."[13] In cyberspace, one may have any number of different virtual alter egos operating in a variety of different MUDs, literally *acting out* different cyber-selves in each fantasy domain. In *Computers as Theater* virtual reality researcher Brenda Laurel has indeed drawn a parallel between computer games and virtual worlds and the classical power of drama.[14]

Although this imaginative role-play is most pronounced in MUDs, it also takes place in online chat rooms, in USENET groups, and on IRC channels. In all these environments, netizens create digital alter egos—though not usually ones as fantastical as those found in MUDs. As a publicly accessible realm of psychological play, cyberspace is, I suggest, an important social tool. This digital domain provides a place where people around the globe can *collectively* create imaginative "other" worlds and experiences. Within these worlds you can not only express your *own* alter egos, you can participate in a group fantasy that has the richness of texture generated by many imaginations working together.

In this respect MUDs may in fact be seen as a variation on practices that occur in many cultures. In ancient Greek society, for example, drama was not merely entertainment, it also served as a vehicle for collective psychological catharsis. Moreover, in many cultures, drama includes the audience, who also become *participants* in whatever "alternative reality" is being enacted. Take, for example, the famous Passion play of Oberammergau in Germany. Every decade the entire town joins in a collective reenactment of Christ's final days; the event lasts for days and transforms the town along with its inhabitants. One way of looking at MUDs is as collective dramas, where again everyone in the community becomes a "player." Everyone gets a part and a costume—and as many lines as they want.

Even in our technological age, one does not have to resort to cyberspace to participate in collective role-playing "drama."

Dungeons and Dragons, on which MUDs were originally based, is itself a hugely successful role-playing game. Its endless spin-offs—which include medieval and mystery scenarios—provide plenty of nonelectronic opportunities for the creation of fantastical alter egos. So too do battle board games such as the World War I scenario Diplomacy. During the mid-eighties I was intensely involved for most of a year with a Diplomacy group as we battled it out for control of Europe, making and breaking alliances with one another. As Russia, I became obsessed with my part, and I can still remember the pangs that would accompany news of an ally's betrayal; simultaneous of course was the thrill of one's own devious success. For the final move of our yearlong battle, we all dressed in character and assembled for the denouement. Resplendent in a floor-length velvet crinoline and tiara, for that evening I *was* "The Tsarina."

Another kind of nonelectronic collective theater is provided by battle figurine games such as Warhammer, played by millions of men and boys the world over. Instead of becoming a single character, Warhammer players command armies of Wood Elves, Orks, and the like. The games are accompanied by elaborate manuals outlining the history, mythology, psychology, and fighting strategies of the various groups. In any discussion of contemporary collective drama one must also, of course, acknowledge Trekkies, many of whom engage as deeply and obsessively in the world of *Star Trek* as any MUDer. The universe of Kirk, Picard, and Janeway is as vital a "virtual world" as anything found online.

My favorite example of a nonelectronic dramatic alter ego is provided by Bruno Beloff, a computer analyst in Brighton, England. Beloff regularly paints his body like a zebra; then, stark naked except for this coat of black and white stage paint, takes his zebra-self out into public. The zebra's outings include walks along the Brighton Pier, paddles in the ocean, and even visits to the local pub. For Beloff, "being a zebra is a chance to be honest

about who I am, which is a fantastic release."[15] Others find similar release in weekend visits to "pony clubs," where they spend
their days trotting around in harnesses and their nights sleeping in
stables on straw. Theoretically such options are open to us all, but
in practice it is not so easy for zebras on the streets of Manhattan
or in the suburbs of Peoria. Whenever Beloff's zebra-self is out
and about his girlfriend must keep a careful watch for the police—
public nakedness being technically illegal on the Brighton Pier.

Few people have the wherewithal, or courage, to follow
Beloff's example—and many would not even want to—but for
those who do, cyberspace provides a most useful service. Behind
the protective screen of a computer, MUDs open up a space of
psychosocial play to us all—to everyone, that is, who can afford a
personal computer and a monthly Internet connect fee. Within
the sheltered space of the FurryMuck, thousands of people from
around the world abandon themselves to their own animal liberation, donning virtual hooves and wings, baring virtual tooth and
claw, frolicking in bucolic virtual parks, and (well, they *are* being
animals) enjoying liberal doses of virtual rutting. They can do so
here without fear of arrest or the approbation of disapproving parents and friends. What is important is that cyberspace provides a
publicly accessible and safe space for such imaginative play. It literally opens up a new *realm* for people to act out fantasies and try
on alter egos in ways that many of us would not risk doing in the
physical world. That development is to be welcomed, I believe—
though, as we shall see, we must be careful not to get too carried
away with optimism here.

The value of cyber-psychic role-play is perhaps most evident
when considering more down-to-earth examples. Foremost here,
and the one that has garnered most media attention, is cyber
gender-bending. It is no surprise that most MUDers are young
males, yet, says Shannon McRae, a MUD researcher and herself
a MUD wizard, "a surprising number of these young men take the

opportunity to experience social interaction in a female body."[16] While it is all too easy to overstate the subversive power of such experiences, MUDs *can* create a social space in which the flux of gender is more fluid.

Such fluidity can have surprising consequences. Statistically speaking, a female character in a MUD will often turn out to be a man pretending to be a woman. For this reason actual physical women often find their characters harassed to prove they "really" are female. In an arena where females may "really" be males, men cruising for women will often end up partnering not with a woman, but with another man. Since it is not uncommon for such encounters to end in physical gratification — "sometimes with one hand on the keyset, sometimes with two"[17] — this virtual polymorphism suggests that MUD cultures can be more open than society at large. In MUDs, as in most online arenas, it is impossible to tell if your communicants are anything like the characters their textual descriptors suggest.

In the early days of cyberspace several cyber-neighborhoods were rocked by discoveries of men masquerading as women and using this facade as a lure to intimacy with "real" women. They took advantage of the fact that many women will talk intimately with another woman in a way they would not do so with a man. The famous case of "Joan," on the CB channel of CompuServe, highlights how people can "change" gender online. In the mid-1980s, when Joan presented herself to the CompuServe community, she was, she said, a neuropsychologist in her late twenties who had been crippled, disfigured, and left mute by a drunken driver. Despite these appalling injuries, Joan was warm and witty, giving loving support to many in the community. People trusted her quickly, and women especially became intimate with her. Thus many found it shocking when Joan was unmasked as a New York psychiatrist who was not crippled, disfigured, mute, or even female. "Joan" was in fact Alex, a man "who had become obsessed with his own experiments in being treated as a female and participating in female friendships."[18]

Yet what so upset the CompuServe community in the mid 1980s has become routine a decade later. "To me there is no real body," one MUDer told online researcher Mizuko Ito. Online, she continued, "it is how you describe yourself and how you act that makes up the 'real you.'" For her, the "real life" gender of her MUD friends and sexual partners was of little interest. While we certainly must not let ourselves be blinded by false optimism here (the experience of gendered physical bodies *cannot* be completely overridden with a keyboard), nonetheless, there is something positive here. As McRae notes: if online, boys can play at being girls, and gays can play at being straight, and vice versa, then in cyberspace " 'straight' or 'queer,' 'male' or 'female' become unreliable as markers of identity"[19]. The point is that since in cyberspace labels cannot be easily verified, their determining power is reduced. The concept of gender, while not wholly up for grabs, is at least partially decoupled from the rigid restrictions so often foisted on us by the form of our physical bodies. Here is a space that offers, even if only temporarily and in very truncated form, a chance to at least get a glimpse at other ways of being.

MUDs may also serve a genuinely therapeutic role. In her book *Life on the Screen* Turkle describes a number of people who have used MUD personae as proxies in their struggles with very real psychological problems. Robert, a college freshman whose life had been severely disrupted by an alcoholic father, turned to MUDing as an escape from the trauma and chaos of his life, at one point spending more than a hundred and twenty hours a week online, eating and even sleeping at his computer. But things took a more serious turn when he accepted administrative responsibilities in a new MUD that turned out to be the equivalent of a full-time job. Building and running a complex online world is a task requiring considerable administrative skills and through the experience of overseeing the MUD Robert gained a new sense of control in his life. Furthermore, he was able to use the MUD as a place to talk about his personal feelings in a constructive way, thereby facilitating better relationships outside the MUD. As he later told Turkle: "The computer is sort of practice to get into closer relationships with people in real life."[20]

I am reminded here of a kind of therapy popular in the late seventies. Known as "psychodrama," patients would role-play various scenarios about their own and their family's lives. In child abuse therapy also, role-play is commonly used—often the children act out scenarios with dolls or other toys. Of course not all MUD experiences are positive. For some, the doors of digital perception open only to escapist delusion, and even addiction. "When you are putting in seventy or eighty hours a week on your fantasy character," says Howard Rheingold, "you don't have much time left for a healthy social life."[21] Or for much of anything else.

What could be more pathetic than the declaration by one MUDer that "this is more real than my real life"?[22] One friend of mine almost lost his long-term relationship when he became so obsessed with the online world of the LambdaMOO he was spending more time with his friends "there" than with his "real life" love. But in this sense, again, MUDs are not unique. *All* fantastical activities—be it playing Dungeons and Dragons, going to Trekkie conventions, snorting cocaine, or drinking alcohol—are open to abuse. Of course MUDs pose the additional problem that they are readily accessible twenty-four hours a day. As a "drug" they are a most convenient and very cheap option.

Throughout cyberspace—in MUDs and chat rooms, on USENET groups and IRC channels—netizens around the globe are engaging in psychosocial experimentation and play. On any day, at any time, thousands of people the world over are launching psychic test balloons into this new space of being. Many insist that their lives contain a dimension that is *not* physically reducible.

Embodied or not, "cyber-selves" are real, and the space of their action, though immaterial, is nonetheless a genuine part of reality.

This cyberspace-induced dualism can only intensify. As ever more communications media, businesses, newspapers, magazines, shopping malls, college courses, libraries, catalogs, databases, and games go online we will increasingly be forced to spend time in cyberspace—whether we want to or not. My godson, Lucien, is growing up with the Internet; he does not know a world without it. His generation (at least in the industrialized world) will hardly have a choice about whether to participate in this new space. One proleptic example: UCLA recently requested that every one of its undergraduate courses have an accompanying Web site. Whether driven by imperatives to cut costs, or by genuine desire to improve the learning environment, education is just one area that will increasingly move online. For Lucien and his friends, cyberspace will be an unavoidable parallel world that they will *have* to engage with.

Before we get too upset about this bifurcation of reality, it is well to remember that those of us born after the mid-fifties have *already* been living with a collective parallel world—the one on the other side of the television screen. We who grew up with *Bewitched, I Dream of Jeannie, Gilligan's Island,* and *Get Smart*— are we not already participating in a vast "consensual hallucination"? One that, as in *Bewitched,* is deeply imbued with magical qualities (see Figure 6.1). The collective drama of soap operas and sitcoms—be it the daytime fare of *Days of Our Lives* and *General Hospital,* or the nighttime fare of *Melrose Place* and *Seinfeld*—are these not "consensual hallucinations" which engage tens of millions of people around the world every day of the week? What is the cartoon town of Springfield in *The Simpsons* if not a genuine "virtual world"?

It is well to remember also that throughout human history all cultures have had parallel "other" worlds. For Christian medievals,

FIGURE 6.1. The "consensual hallucination" of television has already paved the way for the parallel world of cyberspace.

as we have seen, it was the world of the soul described by Dante. For the ancient Greeks it was the world of the Olympian gods and a host of other immaterial beings—the Fates, the Furies, et cetera. For the Aboriginal people of Australia it was the world of the Dreamtime spirits. And so on. I do not mean to imply here that the Greek gods or the Aboriginal Dreamtime spirits were nothing more substantial than television characters (quite the opposite is true), I only want to point out that a *multileveled reality* is something humans have been living with since the dawn of our species.

With the virtual world of television we in the late twentieth century have once again created another plane of reality, and thereby paved the way for the new dualism of cyberspace. Yet if this dualism between the physical and the virtual worlds is not something entirely new, for our children and their children it will be greatly magnified. As in the Middle Ages, they will increasingly *inhabit* a two-phase reality.

Entering upon this new age of cyber-dualism we may wish to look afresh at the dualism of the Middle Ages. Can we see ourselves reflected in that distant mirror? Though we must be careful not to fall for glib concordances, Barbara Tuchman's study of the parallels between Dante's century and our own is not without resonances for cyberspace.[23] Much like the cyber-domain today, the medieval afterlife served as a collective parallel world of the imagination.

As with MUDs, the medieval afterlife teemed with nonhuman, half-human, and suprahuman life. Think of Dante's Minos, the demonic judge of Hell, or Geryon, that patchwork creature of man, mammal, and serpent who ferries Dante and Virgil down the chasm to the Malebolge. With his chimeric body and his brightly whorled fur he would be right at home in the FurryMuck. And just look at Hieronymus Bosch's visions of Heaven and Hell. On a small canvas Bosch could conjure an entire virtual world populated by an imaginal cast that would be the envy of any MUD wizard. Moreover, like cyberspace, the medieval afterlife was a place where friends, and even love, might be found. As a guide, teacher, and protector in an often bewildering place, Virgil is surely the paradigmatic virtual friend. And what greater model for virtual love than that between Dante and Beatrice?

Whatever else it is, *The Divine Comedy* is also one of the most truly "fabulous" worlds ever conjured in text. On one level it is a *genuine* medieval MUD. The parallels between *The Divine Comedy* and computer-based virtual worlds have indeed been noted by a number of scholars. According to Erik Davis, both "tend toward baroque complexity, contain magical or hyperdimensional operations, and frequently represent their abstractions spatially."[24] As we have seen, *The Divine Comedy* is organized as a multileveled hierarchy: the nine circles of Hell, the nine cornices of Purgatory, and the nine spheres of Heaven. Dante's journey is an ascent up this ladder. So also in many medieval and combat MUDs; players work their way up through multiple layers of expertise. Virtual ascent through a MUD brings one finally into the "transcendent" class of "wizard"—a cyber-equivalent of Dante's heavenly elect?

Davis has pointed out that one of the very first computer-based virtual worlds, the game Adventure, also has resonances with Dante's world. As the first computerized version of Dungeons and Dragons, Adventure directly inspired the development of the first MUDs. The Adventure player's task, rather like Dante's in the Inferno, was to negotiate his or her way through a hazardous underground maze of caves, and out to the light beyond. On the way, one would search for treasures and magical spells, solve puzzles, and kill trolls. Computer industry chronicler Stephen Levy has suggested that Adventure might also be seen as a metaphor for computing itself. During the game, players cracked the code of this virtual world in much the way that a hacker would crack the code of a computer operating system. Cracking hidden codes in virtual worlds is also a major theme in many cyber-fictions, notably Gibson's *Neuromancer* and Neal Stephenson's *Snow Crash*. So too, Dante scholars stress that the virtual world of *The Divine Comedy* is a complex puzzle of subtle hidden codes.

Cracking these codes, deciphering the multiform patterns both in Dante's world and in the poem that describes it, has become a favorite task of Dante scholars, who comprise, in this sense, a kind of medievalist hacker intelligentsia. Over the last century they have uncovered scores of hidden patterns in Dante's prose and in his world. "These range from relatively accessible insights—[such as] the realization that like-numbered cantos in the *Inferno, Purgatorio* and *Paradiso* have important thematic ties—to truly abstruse discoveries about the positions of critical words or rhymes."[25]

Patterns have been found in the spatial arrangement of the three afterlife kingdoms, in the symmetrical arrangement of the dream sequences in Purgatory, in the number of lines in each canto, the distribution of longer and shorter cantos, and so on. Beneath the sublime poetics of *The Divine Comedy* lies a dazzling substructure of hidden codes. In recognition of Dante's supreme skill as a code wizard, researchers at Lucent Technologies currently designing a revolutionary Net-based operating system have named their system "Inferno." They are hoping that as cyberspace becomes the primary source of computing resources, Inferno will become the global standard operating system, usurping Microsoft's DOS and Windows. Thus Bill Gates would, so to speak, be dethroned by Dante.

I have suggested that the new cyber-dualism is a development to be welcomed, yet we would do well to consider carefully what cyberspace does and does not enable. More so even than with most new technologies, there is an enormous amount of hype surrounding cyberspace. I have endorsed the view that cyberspace provides a new space for experimentation with various facets of selfhood, but some cyber-enthusiasts go much further. In *Life on the Screen*, Sherry Turkle proposes that in this postmodern age of cyberspace, the unity of the self is an old-fashioned fiction. According to Turkle, cyberspace provides the opportunity for splitting the self into a radical *multiplicity.*

In discussing the notion of multiple selves Turkle draws on the computer concept of "windows," the software paradigm that enables a computer user to be working on several different kinds of files at once, each one (say a spreadsheet, a word processing document, and a graphics file) constituting a separate "window." "In the daily practice of many computer users," Turkle tells us, "windows have become a powerful metaphor for thinking about the self as a multiple distributed system." She then goes on, and I quote at length, for the passage, I think, is key. In cyberspace, Turkle says:

> The self is no longer simply playing different roles in different settings at different times, something that a person experiences when, for example, she wakes up as a lover, makes breakfast as a mother, and drives to work as a lawyer. The life practice of windows is that of a decentered self that exists in many worlds and plays many roles at the same time. In traditional theater and in role-playing games that take place in physical space, one steps in and out of character; MUDs, in contrast, offer parallel identities, parallel lives. The experience of this parallelism encourages treating on-screen and off-screen lives with a surprising degree of equality. Experiences on the Internet extend the metaphor of windows—now real life itself [as one of her MUD subjects notes] can be "just one more window."[26]

It is certainly true in the late twentieth century that most of us must negotiate different roles in our daily lives. To that extent we are all multifaceted beings. But to suggest, as Turkle does, that cyberspace offers "parallel identities, parallel lives," which are equal to our physical lives and identities is going too far. True multiple personalities, such as the famous case of "Sybil" are deeply traumatized people with major psychological dysfunction. To play at being a singing fish or the opposite sex can indeed be a positive experience, but to believe that these experiences are *equal* to life in the flesh is delusion. Elsewhere in her book, Turkle tells us that "some [MUDers] experience their lives as 'cycling through' between the real world and a series of virtual worlds."[27] For some players, apparently, these cyber-selves become so "real" they question the privileged position of the physical self. As one of her subjects puts it: "Why grant such superior status to the self that has the body when the selves that don't have bodies are able to have different kinds of experiences?"[28]

One answer is that "the self that has the body" *really* dies. If a cyber-self is killed, or even if a host computer crashes and a whole MUD world is obliterated (as happens on occasion), it can always be rebooted, or you can create a new character and start again. That may not be quite the same experience as with a previous character, but it is a far cry from heart-stopping physical death. Moreover, the self with the physical body *really* gets sick, it *really* feels pain, and crucially, it is bound into a social network of other physical selves whom it cannot simply shut out by logging off the system. People *do* sometimes walk away from their physical lives and disappear, but that is rare for precisely the reason that in the physical world we are *physically dependent* on one another for care and support. Social bonds established in cyberspace can be, and often are, deep and powerful, but these "parallel lives" are *not* equivalent to the lives we experience with our physical bodies.

What is perhaps more troubling about such claims, as philosopher Christine Wertheim has pointed out, is that the notion that we can totally *remake* our "selves" online obscures the very significant difficulties of achieving real psychological change.[29] The notion that we can radically *reinvent* ourselves in cyberspace and create whole "parallel identities" suggests that the very concept of selfhood is endlessly malleable and under our control. In Turkle's vision, the self becomes a kind of infinitely flexible psychic plasticine. What such a vision belies is the enormous amount of psychological shaping and forming that is enacted on an individual by his or her upbringing, by his or her society, and by his or her genes. This shaping, much of which occurs when we are very young, cannot generally be overthrown or reengineered except by an enormous amount of psychological hard work. While I believe wholeheartedly that each of us does have the power to change our "selves" profoundly, real self-transformation is extraordinarily difficult—which is why psychotherapy is usually such a long process.

Role-playing at being a squirriloid or a Klingon, whatever its genuine value, is simply not an identity-changing experience. "I"—that is, my "self"—can role-play any number of different personae online and off, but that does not mean I become fragmented. In every one of these situations I am still me, unless I become a true split personality like Sybil, in which case I am likely to be committed. Moreover there is the problem that if we come to really believe that sane people can be split personalities, then how are we going to apportion responsibility? If one of my "alters" commits murder, does that mean "I" am responsible? Who would go on trial? Surely our goal should not be to encourage the idea of self-fragmentation, but rather to learn to better contain paradoxes within the *one* self. Certainly there are parts of me that disagree with one another, but I consider it a sign of my growing maturity that I no longer seek total internal unity on every issue.

Life in the physical body—what MUDers so quaintly refer to as RL (i.e., real life)—is not the totality of *real life*. In our materialistic age, the inner life of mind *has* generally been accorded too secondary a place in our discourse about reality. But in rehabilitating "mind" back into our conception of the real it will not do to make the *opposite* mistake of denying the unique and irreplaceable role of the body. In a sense, all this is just another iteration of the age-old mind-body tension in Western culture. For the past several centuries the body has been decidedly to the fore in our thinking, which is hardly healthy; yet we ought also to be wary of letting the pendulum swing too far back in the other direction. Life in the flesh is *not* "just another window," and we ought strongly to resist efforts to promote it as such.

As I see it, the value of cyberspace is not that it enables us to become multiple selves (a concept that seems pathological), but rather that it encourages a more fluid and expansive vision of the one self. Perhaps this is what Turkle means by a "decentered self"? The point is that if we allow (as I believe we must) that some part of my self "goes" into cyberspace when I log onto a MUD or onto the Net, then we must also acknowledge that some part of my self also "goes" into every letter I write. If you like, my self "leaks out" in the letters and stories that I write, and even in the phone conversations I have. If I carry on a long-term correspondence by the old-fashioned post (as I have been known to do), there is a sense in which the "I" of those letters is also an extension of me. It, too, becomes a kind of virtual alter ego. As Christine Wertheim puts it, even offline "I am extending myself all over the place."

All this is not to deny that cyberspace provides a *new space* for such extensions of self—one that is, moreover, highly public. It is only to point out that the kinds of self-extensions that occur online also take place in our lives offline. To be sure, this is not generally in such dramatic forms as cyberspace allows, but these extensions or extrapolations of self are going on nonetheless.

One question that arises, then, is *where does the self end?* If the self "continues" into cyberspace, then as I say, it also "continues" through the post and over the phone. It becomes almost like a *fluid*, leaking out around us all the time and joining each of us into a vast ocean, or web, of relationships with other leaky selves. In this sense, cyberspace becomes a wonderful metaphor for highlighting and bringing to our attention this crucial aspect of our lives. As Wertheim points out, the Net makes *explicit* a process that is already going on around us all the time, but which we in the modern West too often tend to forget. By bringing into focus the fact that we are all bound into a web of interrelating and fluid selves, the Internet does us an invaluable service.

Another way of looking at this is to say that every one of us "occupies" a "volume" of some kind of "self-space," a space that "encompasses" us as profoundly as the physical space that modern science describes. This collective "self-space," this communal ocean of leaky selves, binds us together as psychosocial beings. I am well aware that in this materialistic age, such an assertion will be greeted with derision in some quarters. Neuroscientists and philosophers such as Daniel Dennett and Paul and Patricia Churchland, who claim that the human mind can be fully explained in terms of materialistic neurological models, will no doubt scoff at any notion of "self-space." But I suggest that something like this is precisely what we *experience* as thinking, emoting beings. Just such an idea is indeed encoded in many religious and mythological systems.

I do not mean to claim here that "self-space" exists *independently* of physical space, as something ontologically separate. Obviously, my "self" only exists because there is a physical body in which it is grounded. At the same time, "I" am not restricted purely to the space of that body. As Descartes recognized, there is a sense in which I am first and foremost an immaterial being. After three hundred years of physicalism, cyberspace helps to make explicit once more some of the *nonphysical* extensions of human beingness, suggesting again the inherent limitations of a strictly reductionist, materialist conception of reality. Again, it challenges us to look beyond physicalist dogma to a more complex and nuanced conception both of ourselves, and of the world around us.

CHAPTER SEVEN

CYBER SOUL-SPACE

Let us begin with the object of desire. It exists, it has existed for all of time, and will continue eternally. It has held the attention of all mystics and witches and hackers for all time. It is the Graal. The mythology of the Sangraal—the Holy Grail—is the archetype of the revealed illumination withdrawn. The revelation of the graal is always a personal and unique experience. . . . I know—because I have heard it countless times from many people across the world—that this moment of revelation is the common element in our experience as a community. The graal is our firm foundation.[1]

This statement would probably seem at first glance an expression of religious faith. With its focus on the Holy Grail, surely the "community" referred to must be Christian. The clue that it is not is in the second sentence. What is the word "hackers" doing there? In fact this is not an extract from a Christian revival meeting but from the capstone speech to a conference of cyberspace and virtual reality developers. The speech was given by Mark Pesce, codeveloper of VRML, Virtual Reality Modeling Language. With VRML, Pesce is a key force behind the technology that is enabling online worlds to be rendered graphically, thereby moving us closer to Gibson's original cyberspace vision.

As one whose work contributes centrally to the visual realization of cyberspace, Pesce is a man of considerable influence within the community of cyberspace builders and technicians. His views here command attention. What, then, is the "moment of revelation" he declares as "the common element in our experience of a community"? Just what is the almost mystical experience these cyber-architects apparently share?

According to Pesce, for each individual it takes a unique form. For him personally it took the form of a William Gibson short story—an early precursor to *Neuromancer*. True to the mythic archetype he outlines in his speech, Pesce told his cyber-colleagues the experience occurred at a time of crisis: He had just been expelled from MIT, and was on his way home by bus to break the news to his parents. To while away those Greyhound hours he had purchased a copy of *OMNI* magazine, wherein he discovered Gibson's seminal foray into cyberpunk. As he recounted, the story "left me dazzled with its brilliance, drenched in sweat, entirely seduced. For here, spelled out in the first paragraph in the nonsense word 'cyberspace,' I had discovered numinous beauty; here in the visible architecture of reason, was truth."[2] Everything that comes after this, he continued, even "our appearance here today—is simply the methodical search to recover a vision of an object that declares its existence outside of time."

The almost ecstatic religiosity underlying Pesce's account of his initiation into cyberspace, and his insistence that such experiences are "the common element" binding together the community of its designers, alerts us to the final phase in the history of space that I want to explore in this book. We have just seen how cyberspace is being claimed as a realm for the "self"; in this chapter we explore how it is also being claimed as a new kind of spiritual space — and even as a realm for the "soul." Thus, as promised in the title of this work, we find ourselves at last before the pearly gates of cyberspace.

In one form or another, a "religious" attitude has been voiced by almost all the leading champions of cyberspace. *Wired* magazine's Kevin Kelly is by no means alone in seeing "soul-data" in silicon. VR guru Jaron Lanier has remarked that "I see the Internet as a syncretic version of Christian ritual, I really do. There's this sensibility and transcendence that's applied to computers, regularly. Where did that come from? That's a Christian idea."[3] Speaking at another conference Pesce has said that it "seems reasonable to assume that people will want to worship" in cyberspace. Elsewhere he refers to "the divine parts of ourselves, that we invoke in that space."[4]

And let us not forget VR animator Nicole Stenger's claim that "on the other side of our data gloves . . . we will all become angels." Like Pesce, Stenger too experienced a moment of quasi-mystical revelation that precipitated her into the cyberspace profession. For her, this occurred while watching an early work of computer animation. Describing the experience, she writes that the animator "had revealed a state of grace to us, tapped a wavelength where image, music, language and love were pulsing in one harmony."[5] According to Stenger, those "who decided to follow the light" would "find a common thread running through cyberspace, dream, hallucination, and mysticism."[6]

In some cyber-fiction, cyberspace itself becomes a kind of divine entity. In the *Neuromancer* sequel *Mona Lisa Overdrive*, one of the superhuman artificial intelligences who inhabits the novel's cyberspace explains that the "matrix" (i.e., the Net) exhibits qualities of omniscience and omnipotence. Is the matrix God? asks one bemused human. No, we are told, but you could say that "the matrix has a God."[7] As anthropologist David Thomas has noted, Gibson's novel suggests "that cyberspace must be understood not only in narrowly socioeconomic terms, or in terms of a conventional parallel culture, but also . . . [as a] potential creative cybernetic godhead."[8]

Whether or not the champions of cyberspace are formal religious believers (like Kevin Kelly), again and again we find in their discussions of the digital domain a "religious valorization" of this realm, to use the apposite phrase of religious scholar Mircea Eliade. Claims such as Stenger's that "cyberspace will feel like Paradise," call to mind Eliade's notion that even in secular societies "man . . . never succeeds in completely doing away with religious behavior."[9] Whether or not that is true for "man" in general, it certainly seems close to the mark for cybernautic man and woman.

This projection of essentially religious dreams onto cyberspace is not, as I have already suggested, particularly surprising. As a new immaterial space, cyberspace makes an almost irresistible target for such longings. From both our Greek and our Judeo-Christian heritage Western culture has within it a deep current of dualism that has *always* associated immateriality with spiritually. Stenger herself explains how cyberspace fits this pattern. In the anthology *Cyberspace: First Steps*, she cites with approval Eliade's view that for religious people space is not homogeneous, but is divided into distinct realms of "profane space" and "sacred space." According to Stenger, because cyberspace is a different kind of space from the "profane space" of the physical world, then it "definitely qualifies for Eliade's vision" of sacred space.[10] She argues that indeed cyberspace creates "the ideal conditions" for what Eliade terms a *"heirophany"*—that is, "an irruption of the sacred."[11]

Throughout history, in cultures across the globe, religion has played a central role in people's lives, and as the current tsunami of religious and quasi-religious enthusiasm attests, the desire for a "spiritual life" continues in America today. Through prayer, meditation, retreats, home churching, spirit chanelling, and psychotropic drugs, people across the nation are seeking pearly gates of one sort or another. And the United States is far from alone in this trend: Around the world, from Iran to Japan, religious fervor is on the rise. In this climate, the timing for something like cyberspace could hardly have been better. It was perhaps inevitable that the appearance of a new immaterial space would precipitate a flood of techno-spiritual dreaming. That this site of religious expectation is being realized through the by-products of science—the force that so effectively annihilated the soul-space of the medieval world picture—is surely one of the greater ironies of our times.

Speaking of the dreams that people project onto science and technology, philosopher Mary Midgley has written that "Attending to the workings of the scientific imagination is not a soft option. [This imagining] is not just harmless, licensed amusement. It plays a part in shaping the world-pictures that determine our standards of thought—the standards by which we judge what is possible and plausible."[12] As a subset of the scientific imagination, the cyber-imagination is becoming a powerful force in shaping our world, and we would do well to attend closely to its workings. What then are the particular forms of cyber-religiosity? What are the specific ideals these techno-spiritualists are beginning to judge as "possible and plausible"? Finally, what are we to make of all this? What does it all mean? These are the question we must ask.

True to the title of this volume, religious dreaming about cyberspace begins with the vision of the Heavenly City—that transcendent polis whose entrance is the legendary pearly gates. A connection between cyberspace and the New Jerusalem has been spelled out explicitly by commentator Michael Benedikt. Benedikt explains that the New Jerusalem, like the Garden of Eden, is a place where man will walk in the fullness of God's grace, but "Where Eden (before the Fall) stands for our state of innocence, indeed ignorance, the Heavenly City stands for our state of wisdom and knowledge."[13] The New Jerusalem, then, is a place of *knowing*, a space that like cyberspace Benedikt says is rooted in *information*.

In the book of Revelation, this key feature of the Heavenly City is signaled by its highly structured geometry, which is glimpsed in the repeated use of twelves and fours and sevens in its description. In this sense the Heavenly City suggests a glittering numerological puzzle, which in contrast to the wilderness of Eden is rigor and order incarnate. It is "laid out like a beautiful equation," Benedikt says. According to him, the Heavenly City is indeed nothing less than "a religious vision of cyberspace."[14] While Benedikt sees the New Jerusalem as a Christian prevision of cyberspace, reciprocally he suggests that cyberspace could be a digital version of the Heavenly City.[15]

On a purely visual level the most famous description of cyberspace—in Gibson's *Neuromancer*—does indeed bear an uncanny resemblance to the biblical Heavenly City. Here too we find a realm of geometry and light that is "sparkling, insubstantial," and "laid out like a beautiful equation." Here too is a glittering "city" adorned with "jewels"—the great corporate databases that decorate the "matrix" with a sparkling array of blue pyramids, green cubes, and pink rhomboids. Built from pure data, here is an idealized polis of crystalline order and mathematical rigor.

Most prominently, the Christian vision of the Heavenly City is a dream about transcendence. Transcendence over earthly squalor and chaos, and above all transcendence over the limitations of the body. For the elect in Heaven, Revelation 21:4 tells us, "God himself . . . will wipe away every tear from their eyes, and death shall be no more; neither shall there be mourning for crying nor pain any more, for the former things have passed away."

In Heaven we are promised that the "sins of the flesh" will be erased and men shall be like angels. Among many champions of cyberspace we also find a yearning for transcendence over the limitations of the body. Here too we witness a longing for the annihilation of pain, restriction, and even death.

Throughout Gibson's cyberpunk novels the body is disparaged as "meat," its prison-like nature contrasted with the limitless freedom that console cowboys enjoy in the infinite space of the matrix. Like the biblical Adam, *Neuromancer*'s hero Case experiences his banishment from cyberspace and his subsequent "imprisonment" in his flesh, as "the Fall." From Lanier's claim that "this technology has the promise of transcending the body" to Moravec's hopes for a future in which we will "be freed from bondage to a material body," the discourse about cyberspace thrums with what Arthur Kroker has dubbed "the will to virtuality."

Dreaming of a day when we will be able to download ourselves into computers, Stenger has imagined that in cyberspace we will create virtual doppelgangers who will remain youthful and gorgeous forever. Unlike our physical bodies, these cyberspatial simulacra will not age, they will not get sick, they will not get wrinkled or tired. According to Stenger, the "eternal present [of cyberspace] will be seen as a Fountain of Youth, where you will bathe and refresh yourself into a sparkling juvenile."[16] As we are "re-sourced" in cyberspace, Stenger suggests, we will all acquire the "habit of perfection."[17]

Nothing epitomizes the cybernautic desire to transcend the body's limitations more than the fantasy of abandoning the flesh completely by downloading oneself to *cyber-immortality*. At the end of *Neuromancer*, a virtual version of Case is fed into the matrix to live forever in a little cyber-paradise. A similar fate awaits Gibson's next hero, Bobby Newmark, who at the end of *Mona Lisa Overdrive* is also downloaded to digital eternity. The dream of cyber-immortality was presaged in what is now recognized as the first cyber-fiction classic, Vernor Vinge's novella *True Names* (published in 1981, three years before Gibson coined the word "cyberspace"). At the end of Vinge's story, the physical woman behind the cyber-heroine, known online as "the red witch Erythrina," is gradually transferring her personality into a cyberspace construct. "Every time I'm there," she explains, "I transfer a little more of myself. The kernel is growing into a true Erythrina, who is also truly me."[18] A "me" that will "live" on forever in cyberspace after the physical woman dies.

Yet there is a paradox at work behind these dreams. Even though many cyberspace enthusiasts long to escape the limitations of the body, most also cling to the glories of physical incarnation. They may not like bodily finitude, especially the part about death, but at the same time they desire the sensations and the thrills of the flesh. In Case's tropical cyber-paradise, he relishes the warmth of the sun on his back and the feel of sand squishing beneath his feet. Above all, he delights in the ecstasy of sex with his cyber-girlfriend Linda Lee. He might not take his flesh into cyberspace, but Gibson's hero is vouchsafed the full complement of bodily pleasures.

Cybernauts' ambivalent regard for the body is indicated by the very metaphor of "surfing" they have chosen. Who more than a surfer revels in the unique joy of bodily incarnation?

Commenting on this paradox, Steven Whittaker has described the typical cyberspace enthusiast as "someone who desires embodiment and disembodiment in the same instant. His ideal machine would address itself to his senses, yet free him from his body. His is a vision which loves sensorial possibility while hating bodily limits."[19] In other words, he wants his cake and to eat it too — to enjoy the pleasures of the physical body, but without any of its weaknesses or restrictions.

Yet is this not also the promise of Christian eschatology? Repackaged in digital garb, this is the dream of the "glorified body" that the heavenly elect can look forward to when Judgment Day comes. As we saw in Chapter One, Christ's resurrection has always been interpreted by orthodox theologians to mean that when the last trumpet sounds the virtuous will be resurrected in *body* as well as soul. "The person is not the soul" alone, wrote Saint Bonaventure, "it is a composite. Thus it is established that [the person in Heaven] must be there as a composite, that is, of soul and body."[20] In the eternal bliss of the Empyrean the elect will be reunited with their material selves to experience again the joy of their incarnated form. But this heavenly body will be a "glorified body," free from the limitations of the mortal flesh. In the words of the medieval scholar Peter the Venerable, it will be a body that is "in every sense incorruptible."[21] Just what it meant to have a body in a place that was, strictly speaking, *outside* space and time was a question that much vexed medieval scholars, but that was indeed the position on which all the great theologians insisted.

Medieval scholar Jeffrey Fisher has noted the parallels between this Christian vision and that of many cyberspace enthusiasts. Just as the Christian body returns in glorified form, so Fisher explains that in contemporary cybernautic dreaming the "body returns in a hypercoporeal synthesis."[22] "Hypercorporeal" because like the glorified body of Christianity, this longed-for "cybernautic body" is not apparently bound by any physical limitations. Like the heavenly Christian body, it too is seen as incorruptible, and ultimately indestructible. Fisher cites, for example, the fact that in many hack-and-slash MUDs players who have been killed can simply reboot themselves. Get your head kicked off? No problem, just boot up another. Transcending the limits of the physical body, this cybernautic body has powers far beyond mortal means. "No longer restricted to what it could see with its bodily eyes or do with its bodily arms, the hypercoporeal simulacrum finds itself capable of amazing feats of knowledge and endurance."[23]

A paradigmatic example of this fantasy occurs in Vinge's *True Names*, which is often cited as an inspiration for real-life MUDs and virtual realities. At the end of the story, in a climactic battle for control of cyberspace, "the red witch Erythrina" and the novella's cyber-hero "Mr. Slippery" succeed in defeating the evil enemy by harnessing the power of the world's telecommunications networks. Millions of channels of data come flooding into their brains. From this vantage, Mr. Slippery can now survey the earth with super-human perception: "No sparrow could fall without his knowledge . . . no check could be cashed without his noticing . . . more than three hundred million lives swept before what his senses had become."[24] He has transcended his mortal coil, Vinge tells us. "The human that had been Mr. Slippery was an insect wandering the cathedral his mind had become." Omniscient, and increasingly omnipotent, as he mind-melds with the entire global network, Mr. Slippery is now Fisher's glorified cyber-self who "can go everywhere and see everything in the total presence of the online database."

Such cybernautic dreams of transcending bodily limitations have been fueled by a fundamental philosophical shift of recent years: The growing view that man is defined not by the atoms of his body, but by an information code. This is the belief that our essence lies not in our matter, but in a *pattern of data*. The ease with which many cyber-fiction writers shuttle their characters in and out of cyberspace is premised on a belief that at the core a human being is an array of data. While atoms can only construct the physical body, according to this new view *data* can construct both body and mind. Indeed, many cyber-fantasies imply that in the end we will not need physical bodies at all, for we will be able to reconstruct ourselves totally in cyberspace. As long as these cyber-constructs are sufficiently detailed, Gibson et al. imply, the illusion of incarnation will be indistinguishable from the real material thing.

Look at Case's girlfriend Linda Lee. Halfway through the novel Lee is murdered, but just before she dies she is uploaded into the matrix for complete simulation in cyberspace. So perfect is this reconstruction, so "real" is her cyber-presence as both a mind and a body that she does not know she is only a digital simulation. Cyber-fiction is full of stories of humans being downloaded, uploaded, and off-loaded into cyberspace. Like the medieval Christian Heaven, cyberspace becomes in these tales a place *outside* space and time, a place where the body can somehow be reconstituted in all its glory. Again, it is not at all clear what it would mean to have a "body" in the immaterial domain of cyberspace, but that is the dream many cyber-enthusiasts are beginning to envision. What is extraordinary here is that while the concept of transcending bodily limitation was once seen as *theologically possible,* now it is increasingly conceived as *technologically feasible.* To quote N. Katherine Hayles, "perhaps not since the Middle Ages has the fantasy of leaving the body behind been so widely dispersed through the population, and never has it been so strongly linked with existing technologies."[25]

Lest one imagine that fantasies of cyber-immortality are just in the minds of science fiction writers, we should note that much of the underlying philosophy guiding this fiction is emerging from the realm of science, from fields such as cognitive science, robotics, and information theory. It's all part of the same imaginative flux that produces the dream of "artificial intelligence." What is human mental activity, these believers say, but a pattern of electrical signals in a network of neurons? Why should such a pattern not also be constructed in silicon? AI advocates insist that if computers can be "taught" to do such tasks as parsing sentences and playing grandmaster chess, it should only be a matter of time before they will be able to simulate the full complement of human mental activity.

In the futuristic worlds of many cyberpunk novels this goal

has of course been realized. Gibson's matrix, for example, is inhabited by a slew of superhuman AIs; the eponymous Neuromancer is one of them. Far more than mere calculators, these computer constructs are personalities with their own emotions, desires, and egomaniacal goals. From the vision of creating an *artificial mind* inside a computer it is but a short step to imagining that a *human mind* also might be made to function inside a machine. If both types of "mind" are ultimately just patterns of data encoded in electrical signals, then why should we not be able to transfer one from *wetware* to *hardware?* So goes the argument.

This is precisely the fantasy touted by Carnegie Mellon's Hans Moravec, a world-renowned robotics expert. Moravec, whose lab develops sophisticated robots with three-dimensional vision and mapping capabilities, has seriously suggested that digital mind-downloading will one day be possible. In an extraordinary passage in his book *Mind Children,* he imagines a scenario in which "a robot brain surgeon" gradually transfers a human mind into a waiting computer.[26] As you lie there fully conscious, Moravec describes how a robot surgeon would "open your brain case" and begin downloading your mind layer by layer using "high-resolution magnetic resonance measurements" and "arrays of magnetic and electric antennas." Gradually, as your brain is destroyed, your "real" self—that is, your mind—would be transformed into a digital construct. Just how this is all supposed to happen is never really explained; but it is not the details that concern us, it is the overall fantasy.

Moravec is by no means the only scientist thinking along these lines. The respected mathematician and computer scientist Rudy Rucker has also envisaged downloading human minds to computers in his novels *Wetware* and *Software.* Another real-life champion of the mind-download is Mike Kelly, a Ph.D. in computer science and founder of the Extropian movement. Extropians give even science fiction writers a run for their money, because their goal is ultimately immortality in *physical form*. As Woody Allen once quipped: "I don't want to achieve immortality through my work, I want to achieve it by not dying." Extropians imagine eternal life becoming possible through a powerful cocktail of new technologies, ranging from genetic engineering to nanomachines capable of repairing individual cells. But as they wait for the day when their bodies can be immortalized, Kelly has suggested that they should download their minds into computers as a sort of cyber–waiting room for the main event. Like Kelly, most Extropians are technologically literate young men and women, many employed in fields such as computer science, neurobiology, and even rocket design. Among their heroes are Vinge and Moravec; Moravec himself gave the keynote address at their inaugural conference.

According to many cyber-immortalists, even if there was a catastrophic systems crash you wouldn't necessarily "die," because you could always be restored from backup files kept offline. As in the New Jerusalem, "death would be no more." Moravec himself foresees such a future. In *Mind Children* he writes breathlessly about the day when we will all have backup copies of ourselves stored on computer tape. "Should you die," he says, "an active copy made from the tape could resume your life."[27] True, there would be a bit of a gap between the time when the last backup copy had been made and the moment of your cyber-death, but according to Moravec, "a small patch of amnesia is a trivial affair compared with the total loss of memory and function that results from death in the absence of a copy."

Immortality, transcendence, omniscience—these are dreams beginning to awaken in the cyber-religious imagination. To paraphrase Midgley, these are the things some cyberspace enthusiasts are beginning to think of as "possible and plausible." Myself, I cannot imagine a worse fate than being downloaded into immortality in cyberspace. In Christianity, the elect are promised an eternity of bliss in the presence of ultimate Grace, but what would be the fate of an immortal cyber-elect? What would one *do* in cyber-eternity? There are only so many times you can read the complete works of Dante or Shakespeare or Einstein, there are only a finite number of languages to learn; and after that eternity is *still* forever.

But even for those who desire cyber-immortality there is a fundamental problem: Is the human mind something that could, even in principle, be downloaded into a computer? Is it something that could *ever* be reconstructed in cyberspace? Most cyber-fiction writers and scientists like Moravec assume a priori that since the human mind is an emergent property of the human brain, then it must be just a pattern of electrical data—hence it must ultimately be possible to transfer the files, as it were, from a brain to a computer. This is a similar software metaphor to the one we saw at work in the previous chapter with the idea of the "self" as a set of computer "windows." But is the human mind really just a pattern of data, a collection of "files" that could be transferred from one physical platform to another?

One reason such a vision is problematic is that a human mind has *faulty* memories, and even entirely *false* ones. Memory researchers have shown that by the time we reach adulthood most of us "remember" things that never actually happened; our brains have somehow "created" events that to us seem entirely real. How would such false memories be programmed into a computer? How, in effect, would a machine be taught self-delusion? Moreover, in a human mind many memories are buried well below conscious awareness, yet if they are properly triggered, these memories come flooding back. How would a computer know which memories were supposed to be the conscious ones, and which were to remain unconscious? How would it know at what level of activation the unconscious memories should be triggered into consciousness? How would it have an "unconscious" at all?

In Moravec's scenario, a human mind would supposedly be downloaded in situ. Here, each "layer" of the physical brain would be recorded in sequence into a computer in one continuous sitting. According to him, such a process could capture the *whole mind* in one go. But again we have the problem of unconsciousness. Let us imagine, for argument's sake, that Moravec's setup was possible, and that you could completely record the set of electrical signals going on in someone's mind at some particular time. Now at any moment a human mind can only be recalling a finite range of thoughts and memories. It cannot be thinking about *everything* it has ever known. How could Moravec's process possibly capture the complete range of memories and knowledge that were not remotely within conscious awareness at the time of the recording?

For Moravec's process to work you would have to argue that *every* memory and *every* piece of knowledge that someone possessed were somehow electrically present at every waking moment. But in that case every moment would be one of *omniscience*. I find such a notion untenable. One of our most fundamental experiences as conscious beings is of time passing, precisely because every moment is *different*. The human mind is quintessentially a *dynamic* phenomena, and it seems absurd to suggest that you could capture it "all" at any one moment. I do not raise these objections to be churlish, but only to point out the degree to which mind-download fantasies elide over very real difficulties, not merely with respect to the technology, but more importantly with respect to the perplexing question of just what is a human mind.

By far the most bizarre aspect of mind-download fantasies is the dream of *cyber-resurrection*—the notion of reconstructing in cyberspace people who have died. At the start of Gibson's *Count Zero*, a mercenary named Turner has just been blown to pieces by a bomb. While he waits for the medics to grow him a new body, Turner "himself" (that is, his mind) spends his time in a virtual reality simulation of a nineteenth-century childhood. When his new body is ready his mind will be downloaded into it; in the meantime the otherwise dead Turner whiles away his time in cyberspace. Moravec too dreams of cyber-resurrection, but he goes even further, for he suggests that as a species we may be able to defeat death entirely. He asks us to imagine a brace of "superintelligent archaeologists armed with wonder-instruments." According to Moravec, these digital miracle workers should be able to perfect a process whereby "long-dead people can be resurrected in near-perfect detail at any stage of their life."[28] These undead would be brought back to life in a vast computer simulation. As Moravec writes, "wholesale resurrection may be possible through the use of immense simulators." For Christians, resurrection is promised when the Last Judgment comes, but if Moravec's vision of the future is correct we can expect it well before then.

What we have here, with these visions of cyber-immortality and cyber-resurrection, is an attempt to re-envision a *soul* in digital form. The idea that the "essence" of a person can be separated from his or her body and transformed into the ephemeral media of computer code is a clear repudiation of the materialist view that man is made of matter alone. When the further claim is made that this immaterial self can survive the death of the body and "live on" forever beyond physical space and time, we are back in the realm of medieval Christian dualism. Once again, then, we see in the discourse about cyberspace a return to dualism, a return to a belief that man is a bipolar being consisting of a mortal material body and an immaterial "essence" that is potentially immortal. This posited immortal self, this thing that can supposedly live on in the digital domain after our bodies die, this I dub the "cyber-soul."

It is an astonishing concept to find emerging from the realm of science and technology, but again I suggest this is not wholly an unexpected development. This posited cyber-soul may indeed be seen as the culmination of a tradition that has been informing Western science for over two thousand years. I refer to that curious admixture of mathematics and mysticism that traces its origins to the sixth century B.C. and the enigmatic Greek philosopher Pythagoras of Samos. Whether they realize it or not, today's champions of mind-download not only follow in a Christian tradition, they are also heirs of the Samian master.

As the man who is credited with introducing the Greeks to mathematics, Pythagoras was one of the founders of the Western scientific enterprise. At the same time he was a religious fanatic who managed to fuse mathematics and mysticism into one of the most intriguing syntheses in intellectual history. A contemporary of the Buddha in India, of Zoroaster in Persia, of Confucius and Lao-tzu in China, Pythagoras was a mystic of a uniquely Western stripe. Half a millennia before the birth of Christ, he formulated a radically dualistic philosophy of nature that continues to echo in cybernautic visions today. According to the Samian sage, the essence of reality lay not in matter—in the four elements of earth, air, fire, and water—but in the immaterial magic of *numbers*. For Pythagoras, the numbers were literally gods, and he associated them with the gods of the Greek pantheon. True reality, according to him, was not the plane of matter, but the transcendent realm of these number-gods.

For Pythagoras, the soul too was essentially mathematical. To him it was the soul's ability to express things rationally—literally in terms of *ratios*—that was its primary characteristic. In Pythagorean cosmology, the true home of the soul was the transcendent realm of the number-gods, and after death this is where all souls would return. Unfortunately, during our mortal lives this immortal spark is trapped within the prison of the body, from which it longs to be freed. For the Pythagorean, the aim of religious practice—which necessarily included the study of mathematics—was to free the soul from the shackles of the flesh that it might ascend as often as possible into the divine mathematical realm beyond the material plane.

Even from this cursory description, we see immediately the Pythagorean undertones in contemporary cybernautic dreams. Whatever is downloaded into computers must necessarily be expressed in terms of numbers—to be precise, in terms of the numbers "zero" and "one." The sublimely simple yet infinitely malleable code of zeros and ones is the erector set from which all cyberspace constructs are built. Behind dreams of mind-download is thus a profoundly Pythagorean attitude. Like the ancient Pythagoreans, today's mind-download champions see the "essence" of man as something that is numerically reducible; like the Pythagorean soul their "cyber-soul" is ultimately mathematical. This cyber-soul's "true" home is not the realm of the "meat," but the "eternal" domain of digital data. We have here, then, what Eliade would call a "crypto-religion," a quasi-religious system in which cyberspace reprises the role once accorded the divine space of the ancient Pythagorean number-gods.

Parallels between ancient Pythagoreanism and the new cyber-Pythagoreanism go even further. One of the central beliefs of ancient Pythagoreanism was the eternal return of the soul, a doctrine some believe Pythagoras took from India. Like Hindus, the Samian master believed the soul was continually reincarnated in a series of physical bodies. It was between these incarnations that it bided its time in the realm of the number-gods. A similar process of *metempsychosis* is also featured in a number cyber-fiction fantasies, notably in Rudy Rucker's *Wetware* and *Software*. In these novels, after the main character is uploaded for storage in a central computer, he is periodically downloaded into a series of ever more sophisticated android bodies. As the centuries pass he is reincarnated again and again, his cyber-soul returning each time to the physical world after refreshing respites in a transcendent cyberspatial "Void."

Here Rucker describes this Void, his imagined space of digital disincarnation:

> When you're alive, you think you can't stand the idea of death. You don't want it to stop, the space and the time, the mass and the energy. You don't want it to stop . . . but suppose that it does. It's different then, it's nothing, it's everything, you could call it heaven.[29]

Rucker's Void, his cyber-heaven, is simply a modern updating of the old Pythagorean number-heaven, an eternal space for the "soul" to rest in between its bouts of material incarnation.

But is there not something missing from this technological reincarnation? What about a moral or ethical context? In Hinduism, the form in which one is reincarnated in the next life depends on one's moral choices in past lives. For Hindus, *metempsychosis* is also a moral process. Eventually, in the Hindu scheme, there is supposed to be an end to the process, when one finally attains "enlightenment" and the rounds of reincarnation cease. (In Christianity, where the soul is granted but a single incarnation, there is a much more draconian moral context because it has but *one* chance to make the "right" choices—or pay the price forevermore.)

For ancient Pythagoreans, also, the soul was quintessentially a moral entity. In particular, they believed the soul needed constant cleansing, and they adhered to strict codes of behavior as well as strict regimens of fasting and bodily purification. The cyber-soul, however, has *no* moral context. In cyberspatial fantasies of reincarnation and immortality, the soul's eternity entails no ethical demands, no moral responsibilities. One gets the immortality payoff of a religion, but without any of the obligations. For Pythagoras, who believed that numbers themselves had ethical qualities, the separation of the soul from any moral framework would have been appalling. In the original Pythagoreanism, to take away the moral context would have been to spiritually bankrupt the entire system—which is effectively what the new cyber-Pythagoreanism has done.

For Pythagoras, numbers not only constituted the basis of the divine realm, they also served as the archetypes for the material realm. And here again we see this vision reflected in contemporary cyber-dreams. According to Pythagoras, numbers literally informed the world of matter. He was led to this conclusion by the observation that numbers themselves have forms. As in Figure 7.1, four dots can be arranged in a square, as also can nine dots, or sixteen. Six dots can be arranged in a triangle, as can ten dots, or fifteen. Other numbers make a variety of different forms. If all numbers have forms, Pythagoras reasoned, then might not all forms have number? Might not number be the very essence of form? Twenty-five hundred years later cyberspace is being built on this premise. The very idea of a computer-based model or digital simulation presupposes that form can be captured in the ephemeral dance of number. This is the essence of "virtual reality."

In the cyber-city of AlphaWorld, for example, one can walk down virtual streets lined with virtual trees and flanked by virtual buildings, all of which are ultimately just patterns of zeros and ones residing in a computer memory. As in Figure 7.2, citizens of AlphaWorld can "build" their own virtual homes. Along with reg-

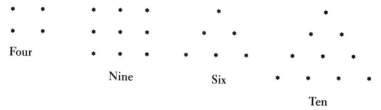

FIGURE 7.1. Pythagoras reasoned that if numbers have forms, then number might be the very essence of form.

FIGURE 7.2. In the cyber-city of AlphaWorld, citizens can build their own virtual homes.

ular houses, people have built pyramids, castles, Greek temples, and myriad other forms. They can even create entire other worlds.[30] Literally "insubstantial"—lacking substance—cyber-forms are nothing *but* patterns of numbers. At the moment most VR worlds are rather crude, but the technology is rapidly improving. Recently I saw an impressively realistic simulation of an Egyptian tomb made by the Italian company Infobyte. Though constructed entirely of ones and zeros there was a convincing illusion of "really being" there in an underground warren of painted rooms and passages. The same company has also made a VR simulation of the Basilica of Assisi, complete with Giotto's *Saint Francis* cycle of frescoes. As evident in Figure 7.3, the illusion of this medieval space is startling in its detail and complexity.

Hans Moravec has suggested that one day we will be able to

FIGURE 7.3. The Italian company Infobyte has created a virtual reality simulation of the medieval Basilica of Assisi.

build in cyberspace a virtual reality model of our entire planet. He imagines

> an immense simulator . . . that can model the whole surface of the earth on an atomic scale and can run time forward and back and produce different plausible outcomes . . . Because of the great detail, this simulator models living things, including humans, in their full complexity.[31]

Where the juices really get flowing here is in Moravec's assertion that with the simulation we could "run time forward and back" and try "different plausible outcomes" for our planet's history. In effect the model would become a digital archetype of our "world"—a purely *numerical form* of planet earth. Whatever else it is, this totalizing cyber-model is a profoundly Pythagorean fantasy.

The Pythagoreans were interested in the numerical forms that inhered in the material world—and in this sense they represent the origin of the science we now know as physics—but first and foremost, Pythagoreanism was a religion. Specifically, it was an ascent religion. Through a strict regimen of bodily and spiritual purification, and by careful study of mathematics, the religion of Pythagoras promised unmediated experience of the All. In other words, *gnosis*—that union with divinity that is characterized by a state of intuitive all-knowingness. For Pythagoras the source of divinity, the fountainhead of the All, was the supreme godhead of the number One, which he associated with the sun god Apollo. The particular form of gnosis that he introduced into the West would reverberate through the ages, infusing the history of science and intersecting continually with Christianity. As we shall see, many cyber-religionists today also aim at an essentially Pythagorean gnosis.

As I have shown in *Pythagoras' Trousers* the Pythagorean spirit was a major catalyst for the emergence of modern physics during the scientific revolution of the seventeenth century—and has remained a force within that science ever since.[32] All the talk from TOE physicists today about "the mind of God" is but a diluted residue of the Samian master's powerful brew of mathematical mysticism. The hyperspace physicist seeking his all-encompassing "theory of everything" is really an updated version of the old Pythagorean sage seeking ultimate knowledge of the One. In both cases what is being sought is the mathematical secret of the All, a numerical form of gnosis.

A similar gnostic spirit can also be discerned among many cyberspace enthusiasts, for here too we find a strong desire for mystical union with the All. Again, Vinge's *True Names* provides a paradigmatic example. When we left Mr. Slippery and Erythrina before, they were on the verge of holding the whole terrestrial world in their minds. By the final moments of their cyberspace battle they have succeeded in this goal. Psychically mind-melding with the entire global network, Mr. Slippery can now see into every place, register every action, follow every transaction taking place on the face of the earth.

Every ship in the seas, every aircraft now making for safe
landing, every one of the loans, the payments, the meals of
an entire race registered clearly on some part of his consciousness. . . . By the analogical rules of the covens, there
was only one valid word for themselves in their present state:
they were gods.[33]

Though the details are numbingly pedestrian, nonetheless
this is a classic description of gnosis, a fusion of the self with the
All that results in a state of *omniscience*. Hildegard of Bingen he
is not, but Mr. Slippery too would be as *one* with the world.

A gnostic drive is also at work in Gibson's *Mona Lisa
Overdrive*, where we read that the "mythform" of cyberspace "in-
volves the assumption of omniscience . . . on the part of the matrix itself." Here, it is the matrix rather than the man that
experiences the All; *it* then communicates this knowledge to its
human interlocutors. According to David Thomas the mythologic
expressed in Gibson's novel "suggests that one of cyberspace's
more fundamental social functions is to serve as a medium to
communicate a form of 'gnosis, mystical knowledge about the nature of things and how they came to be what they are.' "[34] In a
sense Gibson's matrix becomes a digital version of what
Renaissance Neoplatonists called the "world soul," the global intellect which they believed mediated between the human intellect
and the divine intellect.

Real-life cyberspace enthusiasts also have cyber-gnostic
dreams. Hans Moravec's fantasy of a computer model of our entire planet, complete down to the last atom, is but a gnostic longing for godlike knowledge of the All. Like the ancient
Pythagoreans, many of the new cyber-Pythagoreans seek to transcend humanity's humble perspective and acquire omniscient
godlike vision—either of the cyberspatial world, or in the case of
Vinge and Moravec, of our actual physical world.

In contemporary dreams of cyber-gnosis there is a particularly interesting historical parallel with the tradition known as Hermeticism, that during the Renaissance fused together Neoplatonic thinking (itself deeply influenced by Pythagoreanism) and Christianity. Like the ancient Pythagoreans, Renaissance Hermeticists aimed at mystical union with the All.[35] Since, however, they were now operating in a Christian context, that meant fusion with the biblical God. Hermeticists believed that man could not only come to *know* God—i.e., the All—he would become himself like God. The secret, they believed, was to mirror God's relationship with the world. That relationship was summed up in one of their texts by the statement that God "contains within himself like thoughts, the world, himself, the All."[36] To become like God, then—to know the All—the Hermeticist would have to emulate this action and acquire *within his own mind* an *image* of the world. As historian Francis Yates explains, for the Hermeticist "gnosis consists in reflecting the world within the mind."[37]

"Reflection" of the world within the mind is precisely what we see in many cyberspace scenarios. This is what happens at the end of Vinge's novella, where Mr. Slippery and Erythrina finally hold within their minds the whole of cyberspace. By this internal mirroring, we are told, they can now see, and hence control, the whole terrestrial arena. In Gibson's *Neuromancer* trilogy also, there is the implication that the true cognoscenti can contain the whole cyber-domain in their minds at once. Case and the other console cowboys are constantly striving to maximize their internal cyber-vision. On a slightly different tack, in Marc Laidlaw's *Kalifornia*, everyone on earth is (literally) wired into a global network, via "polywires" threaded like silicon nerves through each individual brain and body. Here, a woman known as "the Seer" channels through her mind the mental flux of all humanity. Via the polywire Net she becomes as one with the whole human race.

For Renaissance Hermeticists, the key to obtaining godlike

inner vision was to expand one's sensorium, and ultimately the very boundaries of one's self so that it might encompass all of reality. The above mentioned Hermetic text continued its advice with the exhortation:

> Make yourself grow to a greatness beyond measure, by a bound free yourself from the body; raise yourself above all time, become Eternity; then you will understand God. Believe that nothing is impossible for you, think yourself immortal and capable of understanding all, all arts, all sciences, the nature of every living being.

Reading this passage, one cannot but be struck by how much it resonates both with Vinge's description of Mr. Slippery's powers at the height of his cyberspace battle, and also with Moravec's vision of a total-world cyber-model. Does not Mr. Slippery also make himself "grow to a greatness beyond measure"? Does he too not "by a bound free [him]self from the body"? Does not Moravec, with his totalizing simulation, "believe that nothing is impossible" for him? Even, as we have seen, the ability to control the flow of time—which would of course raise *him* "above all time." Does he too not think himself "capable of understanding all, all arts, all sciences, the nature of every living being"? Surely he would have to in order to simulate the entirety of human culture on the face of the earth. And as for becoming "Eternity," is this not what both men aim for with their visions of cyber-immortality?

There is nothing new about this kind of techno-religious dreaming. During the Renaissance, Hermetic magic was itself seen as a new kind of science, with its practical applications constituting a new kind of technology. Hermeticism *itself* was already then a form of techno-religious dreaming. In the late sixteenth century the great Hermetic practitioner Giordano Bruno wrote that because man has the power, through technology, "to fashion other natures, other courses, other orders" then "he might in the end make himself god of the earth."[38] At around the same time Johann Andreae, very probably the author of the Rosicrucian manifestos, declared that it was man's *duty* to practice the technical arts "in order that the human soul . . . may unfold [itself] through different sorts of machinery." For Andreae, technology provided the means by which "the little spark of divinity remaining in us may shine brightly."[39]

As historian David Noble has shown, in the Christian West champions of technology have been reading religious dreams into technological enterprises ever since the late Middle Ages. If "the technological enterprise . . . remains suffused with religious belief" Noble writes, that is hardly surprising, for "modern technology and religion have evolved together."[40] The pattern of seeing new technology as a means to spiritual transcendence has been repeated so many times that Erik Davis has coined the term "techgnosis" as a generic description of the phenomena.[41] As the latest incarnation of techgnosis, cyber-gnosis reflects a deep and recurring theme in Western culture.

In the glorious futures imagined by cyber-religionists like Vinge and Moravec, godlike omniscience and immortality will be vouchsafed to everyone. *This*, then, is the promise of the "religion" of cyberspace: Through the networked power of silicon we can all become as one with the All. Like Case and Mr. Slippery, we too are promised the power to transcend our mortal coils. Freed from the "prisons" of our bodies by the liberating power of the modem, we too are promised that our "cyber-souls" will soar into the infinite space of the digital ether. There, like Dante in the *Paradiso*, we will supposedly find our way "home" to "Heaven."

We would do well to approach such dreams with our skeptical antennae well tuned, for again there is all too often here an element of moral evasion. Even in its nonelectronic forms Gnosticism has long been problematic. With their focus on transcendence, Gnostics through the ages have often inclined toward a Manichaean repudiation of the body, and along with that has been a tendency to disregard the concerns of the earthly world and earthly communities. Orthodox Christian theologians have long stressed that an essential reason for valuing life in the flesh is that on the physical plane we are bound into physical *communities* to whom we have obligations and responsibilities. Someone who does not value life in the body is less likely to feel obligated to contribute to their physical community: Why bother helping a sick friend if you believe he would be better off dead? Why bother trying to extend life in the flesh if you think it is an evil to be transcended as quickly as possible?

Orthodox Christianity has always affirmed the *value* of the flesh. Humanity was created in body as well as soul, the great medieval theologians asserted, and the duty of the Christian is to live life well in body as well as in spirit. Visions of cyber-gnosis and cyber-immortality are often at heart Manichaean, for we see here as well a strong tendency to devalue life in the flesh. Michael Heim is right when he notes that Gibson's vision of cyberspace evokes a "Gnostic-Platonic-Manichaean contempt for earthy, earthly existence."[42] Too often, cyber-religious dreaming suggests a tendency to abandon responsibility on the earthly plane. Why bother fighting for equal access to education in the physical world if you believe that in cyberspace we can all know everything? Why bother fighting for earthly social justice if you believe that in cyberspace we can all be as gods? What would be the point? Commentator Paulina Borsook has noted that the culture of the Silicon Valley cyber-elite is indeed imbued with a deeply self-serving libertarianism that shuns responsibility toward physical communities. It is a tendency she terms "cyber-selfishness."[43]

Behind the desire for cyber-immortality and cyber-gnosis, there is too often a not insignificant component of cyber-selfishness. Unlike genuine religions that make ethical demands on their followers, cyber-religiosity has no moral precepts. Here, as we have noted, one gets the payoffs of a religion without getting bogged down in reciprocal responsibilities. It is this desire for the personal payoff of a religious system without any of the social demands that I find so troubling. In its quest for bodily transcendence, for immortality, and for union with some posited mystical cyberspatial All, the emerging "religion" of cyberspace rehashes many of the most problematic aspects of Gnostic-Manichaean-Platonist dualism. What is left out here is the element of *community* and one's obligations to the wider *social whole*. Ironically, it is in just this communal aspect that cyberspace may ultimately prove to be of the greatest value.

CHAPTER EIGHT

CYBER-UTOPIA

We have seen the extremes that result from dreams of cyber-transcendence; but there is also a more prosaic, more human side to "heavenly" cyber-dreaming. As noted in the opening chapter, many champions of cyberspace proffer this new digital domain as a realm in which we may realize a better life here on earth. This side of "heavenly" cyber-dreaming is concerned not with escapist visions of immortality and Gnostic omniscience, but more pragmatically with the potential of cyberspace to enhance mortal life. In particular, cyberspace is promoted as a space in which connection and community can be fostered, thereby enriching our lives as *social* beings. In these visions, cyberspace becomes a place for the establishment of idealized communities that transcend the tyrannies of distance and that are free from biases of gender, race, and color. In other words, this is a dream of cyber-utopia.

The promise of utopian community is indeed one of the primary appeals of cyberspace. At a time of widespread social and familial breakdown in the Western world, increasing numbers of people suffer from isolation, loneliness, and alienation. In this climate, says commentator Avital Ronell, "virtual reality, artificial reality, dataspace, or cyberspace are inscriptions of a desire whose

principle symptom can be seen as the absence of community."[1]
The Internet, with its vast global web, beckons us all with a vision
of friendship and the hope of inclusion in a wider social whole.

Howard Rheingold, one of the founders of the WELL (an
early and pioneering online community based out of San
Francisco), is one who believes that cyberspace is already creating
better communities. In his landmark study of online culture, *The
Virtual Community*, Rheingold recalls the utopian prediction of
the legendary cyber-pioneer J. C. Licklider that "life will be hap-
pier for the online individual because the people with whom one
interacts most strongly will be selected more by commonality of interests and goals than by accidents of proximity." Speaking of his
WELL colleagues, Rheingold notes that "my friends and I sometimes believe we are part of the future that Licklider dreamed
about, and we often can attest to the truth of his prediction."[2]
Rheingold is no naif, but he does suggest that cyberspace could
help return us to the practices and ethos of an earlier era. Harking
back to the time before we relinquished our public spaces to corporate developers and the electronic media, he writes that
"Perhaps cyberspace is one of the informal places where people
can rebuild the aspects of community that were lost when the
malt shop became a mall."[3]

High-technology entrepreneur Esther Dyson also believes
that cyberspace can foster the development of more utopian communities.

> The Net offers us a chance to take charge of our own lives
> and to redefine our role as citizens of local communities and
> of a global society. It also hands us the *responsibility* to gov-
> ern ourselves, to think for ourselves, to educate our children,
> to do business honestly, and to work with fellow citizens to
> design rules we want to live by.[4]

According to Dyson, "our common task is to do a better job with the Net than we have done so far in the physical world." Dyson believes that is possible: "Because there will be so much information, so much multimedia, so many options [online] people will learn to value human connection more, and they will look for it on the Net."[5]

For a paradigmatic expression of cyber-utopian optimism we might turn to MIT Media Lab director Nicholas Negroponte. At the end of his book *Being Digital*, Negroponte writes:

> Today, when 20 percent of the world consumes 80 percent of its resources, when a quarter of us have an acceptable standard of living and three-quarters don't, how can this divide possibly come together? While the politicians struggle with the baggage of history, a new generation is emerging from the digital landscape free of many of the old prejudices. These kids are released from the limitation of geographic proximity as the sole basis of friendship, collaboration, play, and neighborhood. Digital technology can be a natural force drawing people into greater world harmony.[6]

Again, David Noble reminds us that there is nothing new about this kind of techno-utopianism. Ever since the sixteenth century champions of technology have been touting it as a key to the creation of more "heavenly" communities. Johann Andreae, for example, envisaged the utopian city of Christianopolis, in which the technical arts were assiduously practiced by all citizens. Like many at the time, Andreae believed the time was nigh for the age of perfection promised by the book of Revelation. The advance of science and technology he saw as essential preparation for this millenial age. Likewise, in the City of the Sun envisaged by the Calabrian heretic Tommaso Campanella, every citizen was to be taught technical skills, "intended to give them the wisdom needed to understand, and to live in harmony with, God's creation."[7] Throughout the sixteenth and seventeenth centuries, utopian visionaries imagined that science and technology could help to precipitate a more perfect era in which Christians would live more harmonious and virtuous lives.

The very word "utopia" derives from the visionary community of the same name imagined by the Englishman Thomas More. Like Francis Bacon's "New Atlantis," More's original utopia was an idealized community located on a remote island, far away from the corrupting influence of a decadent world. In both cases inhabitants had created for themselves a kind of earthly paradise, made possible by their piety, their communal spirit, and crucially, by their devotion to the technical arts.

With these utopian visions we witness the emergence of the idea that man, through his *own efforts*, can create a New Jerusalem here on earth. All these visions were profoundly Christian in intent, inspired, as one commentator has put it, by a "yearning to bring heaven down to earth."[8] Rather than having to wait until the Last Judgement for the advent of a prefect community, Renaissance visionaries suggested that men could create heavenly cities themselves, by their application of science and technology. Technology would thus become a medium for *salvation*. Again and again in the age of science, technology has been viewed as a salvific force, a key to a better, brighter, more just world. Noble and Mary Midgley have both traced this techno-utopian spirit through modern Western culture, where it can be found flourishing today in the NASA space community, in the genetic engineering community, and among advocates of artificial intelligence. But if techno-utopianism is by no means a new phenomenon, among cyberspace enthusiasts it reaches a new crescendo.

MIT's William Mitchell is just one who has championed cyberspace as a potentially utopian realm by drawing a parallel between this digital domain and the agora of ancient Athens.[9] As the center of the original democracy, the agora was the place where Athenian citizens met to discuss ideas for the common good. In this nonhierarchical space all were equal and everyone could express their opinions freely. (Everyone, that is, who qualified as a citizen, which in practice meant about two thousand of the city's most prosperous men.) Mitchell, among others, suggests that cyberspace can again serve as an egalitarian public space.

He points to the fact that in cyberspace we are freed from the normal social markers of physical space, such as suburb names and zip codes. Considering what people pay to live in the 90210 zip code, so that they can formally reside in Beverly Hills, there is no doubt that what Mitchell calls "the geocode" can be a powerfully stratifying force in our perceptions of one another. Whether consciously or not, we *do* often make judgments based on social markers. Saying one lives in the Bronx, for example, is likely to invoke an entirely different set of expectations to saying one lives in Manhattan. As Mitchell writes: "In the standard sort of spatial city, *where* you are frequently tells *who* you are. (And who you are will often determine where you are allowed to be.) Geography is destiny."[10] Online, however, no one knows if you come from Beverly Hills or the backwoods, and they cannot judge you as such. In Mitchell's words, "the Net's despatialization of interaction destroys the geocode's key. [In cyberspace] there is no such thing as a better address, and you cannot attempt to define yourself by being seen in the right places in the right company."

Mitchell goes too far, perhaps, when he says there are no "better addresses" in cyberspace—a prestigious ".edu" address (such as *harvard.edu* or *mit.edu*) carries considerably more cachet online than a CompuServe or America Online address. Yet he is right that cyberspace cuts across many traditional "geocode" boundaries. As a potentially egalitarian arena, consider the following two examples.

In March 1998, Stephen Hawking gave a talk at the White House hosted by the president and the First Lady. Piped to the nation by CNN, the world's most famous living physicist expounded on his ideas about the future of science. In the audience were several Nobel Prize winners and a number of America's leading research scientists, several of whom were invited to ask Hawking questions. But along with these luminaries, questions were also invited from the Internet, and ordinary citizens also took part in the event. One should not make too much of such obvious PR exercises; nonetheless the evening was a small indication of the democratic potential of cyberspace, a potential further illuminated by our second example.

At the Horse Shoe Coffeehouse in San Francisco, Internet access can be obtained at fifty cents for twenty minutes. Around the country, similar venues are springing up, providing public spaces where people who may not have Net access at home can surf the Web, participate in online forums, send and collect e-mail. One San Franciscan who avails himself of the Horse Shoe's facilities is a local squatter named CyberMonk. As one Internet observer has noted, "the combination of real and virtual space afforded by the coffeehouse allows CyberMonk to use it as a living room, telephone, and mailbox."[11] Although in his physical community CyberMonk is marginalized, in cyberspace he becomes an equal member of the digital society. With no fixed abode in the "real" world, in the ephemeral domain of cyberspace he has as "solid" an address and presence as any other netizen.

The notion of an "electronic agora" also underlies the concept of electronic town hall meetings, touted ad infinitum during the 1996 U.S. presidential campaign. Vice President Al Gore in particular would have us believe that cyberspace provides the cure to America's democratic decline. No longer need citizens feel left out of the process of government, say the new cyber-agorians; now via the miracle of the modem *everyone* can be involved in public discussions and communal policy decisions. In cyberspace we would thus realize a true democracy, a dream that (as historically low voter turnouts testify) has so evidently failed in our physical communities.

For young people especially, cyberspace beckons as a place where they might build a better "life." For the first time in several generations, Americans graduating from high school and college are finding they are unlikely to have a higher standard of living than their parents. Most will be lucky to match their parents. With "real life" prospects getting tougher by the year, some young Americans are turning to cyberspace instead. The locus of their dreams, as Sherry Turkle has chronicled, are often MUD worlds. One dispirited twenty-something told Turkle bluntly, "MUDs got me back into the middle class."[12] He did not mean this literally; he was referring only to the online world of his MUD where he and his friends are energetic and productive cyber citizens.

Another of Turkle's subjects, Josh, explained his life in the physical world in the following bleak terms: "I live in a terrible part of town. I see a rat hole of an apartment, I see a dead-end job, I see AIDS."[13] In his MUD world, on the other hand, Josh said: "I see friends, I have something to offer, I see safe sex." There, as an expert at building virtual cafes, he is a respected cyber-entrepreneur. According to Turkle, "MUDs offer Josh a sense of participation in the American dream." He hopes that one day when MUDs become commercial enterprises he will be able to turn his cyber–building skills into a *real* living. For young people like this Turkle notes, MUDs provide "a sense of a middle-class peer group."

MUDs may not be paradise, but for an increasing number of America's youth cyberspace seems a more appealing place than the reality of their physical lives. As a space that is free from middle-class slump, and is immune from the problems of urban decay and social disintegration plaguing so many "real life" communities, cyberspace beckons as a decidedly more utopian domain. On the other side of the modem, these young men and women see a space to meet and date in safety, a place where they can have the kind of power and significance increasingly beyond reach in their physical lives.

Yet for all the optimism of cyber-utopians the digital domain is considerably less "heavenly" than many of its champions would have us believe. While it is true that cyberspace *does* enable interaction between people who would not normally have contact in their physical lives, there are already hints that this social leveling is not as universal as we are often told. In short, it is far from clear that the "pearly gates" of cyberspace are equally open to all.

There is an intriguing historical parallel here that may help to cast light on the future of cyber-utopianism. This may seem a surprising analogy, but literary scholar Brian Connery has shown that many features of the new cyber-utopianism were presaged in the first European coffeehouses of the seventeenth century. Like cyberspace, these early coffeehouses also provided a new social space in which people could mix across class lines, enabling nobles and tradesmen to rub shoulders. Here too, Connery says, the coffeehouses could be seen as "reincarnations" of the classical agora. In this respect they constituted a kind of utopian social experiment, which, like cyberspace, held out the promise of a more equal society for all. In considering cyberspace and its potential as a utopian social space, the history of the coffeehouse offers an illuminating case study.

Within the new coffeehouse culture what mattered most was not wealth or title, but a quick wit and a keen grasp of the latest news. As in cyberspace today, topical information was a key commodity, and after the first newspaper was founded in 1665 coffeehouses become primary places for the public dissemination of news. After the establishment of the penny post in 1680 coffeehouses also became natural locations for delivery of mail. Prior to this, mail had been hand-delivered by porters and was a service available only to the rich. By providing a public venue for dissemination of news and mail, coffeehouses served a similar social function to the Internet today with its online news services and its electronic mail. Indeed, Connery says, these venues "served as laboratories for experimentation" with many of the freedoms that would be enshrined in laws and constitutions later in the century—including "freedom of the press, freedom of association and assembly, freedom of speech."[14]

Yet the genuinely democratizing trends opened up by the coffeehouses would prove short-lived. From the start, dissenters objected to the mixing of classes that occurred there, and in truth there was something challenging about a place where, as one seventeenth-century polemicist put it, "a worthy Lawyer and an errant Pickpocket" could meet on equal footing. But it was not just outside forces that worked against the egalitarian spirit of the coffeehouses; internal forces also would play a role in its demise. It is here, Connery suggests, that the history of the coffeehouse "holds a potential warning for those who dream that the Internet will create utopian discursive communities."[15]

Two forces in particular worked against the new egalitarian spirit: "the reestablishment of authority" and "the institution of exclusivity." Both suggest lessons for cyberspace today. In theory, *anyone* could speak at a coffeehouse discussion—in principle all voices were equal—but in practice most places soon became dominated by the voices of a few, or even just one star speaker. Rather than condemning such behavior, proprietors used these star clients "as a draw for other patrons," a strategy that Connery notes is much the same as "online services [today] who tout the participation of stars from Hollywood or the music industry."[16]

Anyone who has participated on USENET groups knows that all voices are *not* equal, with discussion often dominated by a small cadre of regular vociferous posters. "Newbies" to established newsgroups often get a very chilly reception, and at least one popular newsgroup is famous for its harshly inequitable environment. The case of *alt.folklore.urban*, or AFU as it is known, makes for an interesting example of just how quickly "authority" is indeed being reestablished in cyberspace.

If anywhere in cyberspace ought to be egalitarian, AFU should be. This is a newsgroup devoted to debunking myths and "urban legends." Discussions range over a vast spectrum, from old favorites like alligators in the sewers to reports of high-tech Japanese toilets and rumors about the CIA. As the group's Web site explains, AFU is "a great place to get a reality check on anything that 'a friend' told you, or to compare notes about odd things." Yet despite its populist mission, harsh treatment of newbies by AFU regulars is legendary. Here is one netizen's reaction: "Tell you what scares the shit out of me on the Net, AFU. Now there's a newsgroup to dread. Posting as a newbie there should be one of those (often fatal) moves grouped under the same heading as accidentally shooting yourself through the private parts." AFU regulars pointedly set out to bait newbies with mock postings known as "trolls," a form of mockery that holds up to public ridicule those not conversant with the inner subtleties of the culture. Michele Tepper,[17] herself one of the AFU elite, has pointed out that all social groups need internal rules to maintain group identity; nonetheless she notes that the virulent atmosphere of AFU suggests that equal opportunity of expression is *not* a high priority for this cyberspace community.

In AFU we can also witness the second anti-democratizing trend identified by Connery in coffeehouse culture: "the institution of exclusivity." Already the publicly accessible AFU newsgroup has spun off two exclusive, invitation-only lists. In fact, many newsgroups now have exclusive spin-off lists that are *not* open to the public. Connery tells us that a similar move also occurred in the London coffeehouses, as early as the second decade of the eighteenth century. By that time regular denizens had begun to withdraw from the hoi polloi into exclusive private rooms. Eventually, these select gatherings led to the establishment of private gentlemen's clubs. According to Connery, a similar "development may be inevitable within discussion lists and newsgroups" on the Internet.[18]

It is well to remember that until very recently the digital "agora" was in fact an extremely exclusive place. Up until 1993 (when "browser" software for the World Wide Web first became available), few people outside universities and research settings had access to the Net. Even now there are many people who still cannot afford an appropriate computer and a monthly Internet access fee. And that is true even in rich countries like America. If cyberspace is to become a truly equitable place then we are going to have to face the question of how to ensure that *everyone* has equal access. Not just people who are well-off, but also those who aren't. Moreover, if we are serious about creating some kind of cyber-utopia then the rich developed world is going to have to take seriously the task of making the Internet available to developing countries as well.

One aspect of early coffeehouse culture that was *never* egalitarian was its gender mix. Whatever else may have been in flux, male authority was maintained there, and few women participated in this scene. Cyberspace *is* accessible to women, but how much, really, is the "second sex" welcome? Although the wired world does offer genuine opportunities for women, all is not rosy in this supposed paradise of gender dissolution. Behind the utopian rhetoric, the bits can still pack a hefty sexist bite. Volumes have been written about gender and cyberspace, and it is beyond the scope of this book to give more than a passing glimpse at the subject. But let us consider just one example that I think is particularly illuminating—a case of online sexual harassment.

Few women are more acquainted with this subject than Stephanie Brail. In 1993 Brail was the target of intense online harassment that for several months made her cyber-life hell and finally spilled over into her "real life."[19] The incident began when Brail dared to stand up in support of a young woman whom she thought was being unfairly treated on the USENET group *alt.zines*, a group devoted to discussion of alternative magazine or "zine" culture. The young woman had posted a message requesting to talk about "Riot Grrls" zines—Riot Grrls being a subculture of politically astute young women with punk-rock cultural leanings. Given the nature of the newsgroup, and the fact that "zines" are specifically about *alternatives* to mainstream culture, it was a natural request, but some men on the group vehemently protested. Not only did *they* not want to discuss grrl-culture, they didn't want anyone else on the group to either. One hostile male suggested the young woman start her own group: *alt.grrl.dumbcunts*.

Enraged at this inequity, Brail weighed in with comments defending the young woman's right to speak, comments that, by her own admission, were loud and opinionated. What ensued was a flame war. More insidiously, Brail became the target of online sexual harassment. Soon, "reams of pornographic text detailing gang rapes" were pouring into her mailbox. Yet although she had allies on the original newsgroup, many quickly tired of the flame war and became unsympathetic to her plight. Some even said that by complaining about "Mike" (the harasser), she and her allies were censoring *him*.

Events reached a head when Brail received a message from Mike at a separate private e-mail address. This aggressive stranger had somehow accessed what should have been protected information about her personal life. Chillingly, the message read: *I know you're in Los Angeles. Maybe I can come over and fix your "plumbing."* Now Brail began to fear for her physical safety. The offensive only ended when Mike's guard slipped and Brail was able to sleuth out *his* private e-mail address. After that she never heard from him again.

The story ends well, but happy endings are not all that matter, and the case reveals some rather disturbing undercurrents in cyber-utopia. Brail's case may have been extreme, but online nastiness toward women is not unusual: It is a common reason women give for not wanting to participate in many cyberspace forums. In the face of online harassment women are often told to "just fight back," but that may be easier said than done. As Brail points out, "this is easy advice for a loud-mouthed, college-aged know-it-all who has all the time in the world, but does it apply to real, working women who don't have the time and luxury to 'fight back' against online jerks?" Moreover, why should women *have* to fight back as "the price of admission"? "Men don't usually have to jump through a hoop of sexual innuendo and anti-feminist backlash simply to participate."[20] For many women it is so much easier to just log off.

And that is the primary reason for concern about rampant cyber-misogyny. Under the guise of the First Amendment the cyber-elite has mounted a mantra-like defense of freedom of speech, this supposedly core feature of cyber-utopia. But one has to ask: *Freedom of speech for whom?* Not, apparently, for the young woman who wanted to talk about Riot Grrl zines. And not, apparently, for Brail, speaking in her defense. When women who make postings to *alt.feminism* are called "bitches" by angry young men, is *that* freedom of speech? When, on X-*Files* newsgroups, women are told that their lusty postings in praise of David Duchovny are obscene, is *that* freedom of speech? When, on *Star Trek* newsgroups, women are flamed for expressing dissatisfaction with the female roles in the series, is *that* freedom of speech? "How many women," wonders Brail, "have stopped posting their words because they were sick of constantly being attacked for their opinions?"[21] Thus we must ask, *who* is this cyber-utopia really going to be for?

Women aren't the only ones encountering obstacles in the digital domain. Similar barriers also confront homosexuals, non-whites, and non-Anglos. The heavenly vision of a place where "men of all nations will walk in harmony" is one of the prime fantasies under which cyberspace is being promoted, yet despite many cyberspace enthusiasts' public paeans to pluralism, all cultures are *not* equally welcome in cyberspace. On the contrary, commentator Ziauddin Sardar suggests that what we are seeing is not so much a space for vibrant pluralism but a new form of Western imperialism.

Sardar notes that much of the rhetoric used by cyberspace champions is drawn from the language of colonization. Cyberspace is routinely referred to as a "new continent" or a "new frontier" and its conquest and settlement often compared to the conquest and settlement of the "New World." A typical example comes from Ivan Pope, editor of the British cyberspace magazine *3W*, who described it as "one of those mythical places, like the American West or the African Interior, that excites the passions of explorers and carpetbaggers alike." The headline for a cover story from the San Francisco–based cyberpunk journal *Mondo 2000* declared simply, THE RUSH IS ON! COLONIZING CYBERSPACE.

The theme of colonization is also reflected in a widely quoted document titled "Cyberspace and the American Dream: A Magna Carta for the Knowledge Age," which was put together by right-wing think tank the Progress and Freedom Foundation, and based on the ideas of a group that included Esther Dyson and Alvin Toffler. This cyber Magna Carta states bluntly, "what is happening in cyberspace . . . [calls to mind] the spirit of invention and discovery that led . . . generations of pioneers to tame the

American continent."[22] In a similar vein, the Electronic Frontier Foundation's John Perry Barlow has written that "Columbus was probably the last person to behold so much usable and unclaimed real estate (or unreal estate) as these cybernauts have discovered."[23]

But of course the "real estate" of the Americas *was* claimed. The "taming" of the American West that the writers of the cyber Magna Carta would emulate also entailed the "taming" (and often erasure) of dozens of other cultures. According to Sardar, that is also the hidden danger of cyberspace. Rather than embracing other cultures and their traditions, he suggests that "cyberspace is particularly geared towards the erasure of all non-Western histories." As he explains: "If Columbus, Drake and other swashbuckling heroes of Western civilization were no worse than pioneers of cyberspace, then they [too, by association] must have been a good thing."[24] The implication, Sardar notes, is that the colonized people "should be thankful" for all the "wonderful" technologies the Westerners brought. It is certainly worth asking, as Sardar does, why is it that at a time when colonial frontier metaphors are being so critiqued elsewhere they should be embraced by champions of cyberspace.

Whatever this cyberspatial frontier rhetoric implies about our past, perhaps more insidiously it hints at an *ongoing* cultural imperialism. A frontier, by definition, is a place where things are being formed anew. And newness is exactly what many cyber-enthusiasts prize above all else. For too many of them, history is of little interest, because what *really* matters is the future, a glorious unprecedented future that will supposedly emerge Athena-like from their heads. In such an atmosphere of future-worship, Sardar says, there can be no genuine respect for the traditions of any culture. With the world constantly being formed anew at the digital frontier, traditional ways of thinking and being are all too easily reduced to quaint curiosities: "Other people and their cultures become so many 'models', so many zeros and ones in cyberspace."[25] It is a process that Sardar decries as "the museumization of the world."

On a global scale, moreover, cyberspace provides unprecedented opportunities for "corporations [to] trade gigabytes of information about money and death." Let us never forget the role of the military in the initial development of cyberspace, and their continuing presence at the forefront of this technology. It is not insignificant that the first-ever application of multiuser online virtual reality was for an intercontinental battle simulation.[26] Beyond the military one of the greatest users of cyberspace is the financial industry, and it is already known that billions of crime dollars slosh undetected through the world's computer networks, dissolved into apparent legitimacy by the purifying power of silicon. If, as Sardar and others suggest, "cybercrime is going to be *the* crime of the future," then rather than bringing to mind the New Jerusalem, one might wonder if cyberspace will be more like a new Gomorrah.[27]

Thinking about the potential of cyberspace, we might consider all this in Dantean terms. As a man of the Middle Ages, Dante lived before the time when technology came to be seen as a force for creating a New Jerusalem. In *his* time, human action tended to be associated more with the creation of Hell. One of the most powerful messages of *The Divine Comedy* is that Hell is a place we humans make for ourselves. As I noted in Chapter One, in the medieval cosmos Hell was the space literally *within* the sphere of human activity, and it is no coincidence that Dante placed it inside the earth. Metaphorically speaking Dante's Inferno was the inner space of sick men's minds, a place of vanity, delusion, ego, and self-obsession. The poor souls trapped there were doomed to spend eternity wallowing in the human race's collective psychic disease and excrement.

Now cyberspace too is an inner space of humanity's own making, a space where the vilest sides of human behavior can all too easily effloresce. In the past few years neo-Nazi and skinhead sites have proliferated on the web, while USENET groups make it all the easier for racists and bigots to spread their messages of hatred.[28] Surfing such sites, with their openly violent, antisocial, and antigovernment diatribes, is truly to descend into a new circle of Hell. To say nothing of pornography, for which the Web has undoubtedly been the greatest boon since the invention of photography. As Sardar notes, the underbelly of cyberspace is indeed "a grotesque soup." One is reminded here not so much of Paradise, but, as in Figure 8.1, of the other pole of medieval soul-space. In short, while contemporary exponents of the Renaissance tradition see cyberspace as a potentially heavenly place, harking back to the earlier medieval tradition, there is every potential, if we are not careful, for cyberspace to be less like Heaven and more like Hell.

BEYOND CYBER-UTOPIA

Yet having recognized the inadequacy of much cyber-utopian rhetoric and the not insignificant inequities within many cyberspace communities today, I would like to end this work on a positive theme, for it seems to me that in spite of its problems cyberspace does offer us an essentially positive vision. There is a sense in which I believe it could contribute to our understanding of how to build better communities. I do not want to use the word "utopian" here, because that concept has distinctly Christian undertones, and I want to finish on a note that is less Christocentric, less Eurocentric, and more universal. I want to return here at the end of our story to an idea that was introduced at the end of Chapter Six—the notion of cyberspace as a *network of relationships*. It is this inherently *relational* aspect of cyberspace that I believe can serve as a powerful metaphor for building better communities.

By its very nature, cyberspace draws our attention to something that has been implicitly realized by most myth systems and traditional religions the world over—the way in which human be-

FIGURE 8.1. Detail from Arena Chapel *The Last Judgment*. Might cyberspace become less like Heaven, and more like Hell?

ings are bound together into communities by networks of relationships. Since cyberspace itself is a network of relationships, it *epitomizes* qualities that are fundamental to the creation and sustenance of strong community. This point is crucial, and it warrants our attention.

Whatever people are *doing* in cyberspace, and whatever its *content*, cyberspace itself is a network of relationships in a number of different senses. Firstly, at its underlying material level it consists of a physical network of computers linked together by phone cables, optic fibers, and communications satellites. But along with this physical network, there is also a vast nonphysical network, for many of the relationships that constitute cyberspace are purely logical links, implemented only in software. On both levels, the very essence of cyberspace is *relational:* It is a set of relationships between hardware nodes on the one hand, and on the other hand between software entities such as Web sites and Telnet sites.

On both levels, cyberspace can serve as a metaphor for community, because human communities also are bound together by networks of relationships; the *kinship networks* of our families, the *social networks* of our friends, and the *professional networks* of our work associates.

Within any modern community there are also networks of interrelated businesses, networks of social services, networks of churches, networks of health care providers, and so on. Like cyberspace, these human networks also have both physical and nonphysical components. Health care networks, for example, are comprised of a physical collection of hospital buildings, but they also rely on a network of immaterial links between doctors and specialists who refer patients to one another. Here too, there are both "hardware" and "software" components.

In maintaining the integrity of cyberspace as a globally shared space, the upkeep of reliable network links is crucial. Anyone who has ever experienced difficulties logging onto the Net because of a bad line knows how critically cyberspace depends on good network links. In other words, the strength of cyberspace as a *whole* depends on the maintenance of good connections between the various nodes. Again this is a powerful metaphor for human communities, because the strength of our communities also depends on the maintenance of good "strong" connections between nodes—that is, between individuals and between various social groups. Just like cyberspace, the integrity of human *social space* also depends on the strength and reliability of *our* networks.

Another feature that binds together any human community is the fact that a group of people "inhabit" a common "world"— that is, they share a common vision of reality, or a common "worldview." Central to the creation of a common worldview is a common language, for language is the primary means by which we humans *make sense* of the world around us. What is *real* for any people are those things for which they have words, those concepts and ideas which their language literally *articulates*. There is a sense in which language *creates* the world of any people. Now cyberspace itself is a "world" created by language, a world that actually comes into being through the power of specially designed computer languages. Once again, then, in its very ontology cyberspace serves as a metaphor for processes that are central to the creation of human communities.

The "world-making" power of language has been recognized in the myths and creation stories of cultures and religions the world over. In the Old Testament, for example, we find the famous phrase "In the beginning was the Word." With *these* words the ancient authors of the Hebrew scriptures acknowledge that before language there was, in effect, nothing. The world-making power of language is also recognized by the sophisticated cultures of the Australian Aborigines who traditionally sang songs as they walked across the land, believing that their incantations called the land into being. The entire continent of Australia is crisscrossed by a network of walking tracks known as "song-lines," each associated with a complex cycle of songs. For Aboriginal people, these song-lines formed the basis for a continent-wide navigation system, an elaborately structured network by which they rationally *made sense* of their vast land. In essence, through the power of language, the song-lines gave *structure* to a land that for hundreds of miles at a stretch is almost featureless desert. To put this another way, the song-lines transformed the "emptiness" of the desert into an ordered and structured space. They actually generated the geographic space of Aboriginal life. Moreover, these song-lines also provided a network of relationships between the many different communities that populated the Australian continent.

Cyberspace also is a paradigmatic instance of the "world-making" power of language. At every level of electronic communication within the Internet there are special languages or "protocols" to ensure that all the machines can talk to one another. Cyberspace as a communally shared world would simply not be possible without the immaterial power of language. In addition to various "network protocols," there are also special protocols that determine how written text should be encoded for transmission over the Net, and also for how graphics, sound, and video should be encoded. You cannot in fact do anything in cyberspace without invoking numerous electronic languages and protocols.

More critical even than its hardware, cyberspace is made possible by the ephemeral technology of language. As the great philosopher of space Henri Lefebvre would say, the "production" of cyberspace cannot be reduced solely to its physical components.[29] The irreducibility of cyberspace to its physical substrate is evident in its structure, which, as we have noted, is partly physical and partly not. As William Gibson correctly anticipated in his fiction, the essence of cyberspace is not its material connections but its logical (or linguistic) ones. In the end, cyberspace is not just a physical network, it is above all a logical network.

And again, there is a profoundly *communal* dimension to the "production" of cyberspace, for as a matter of practical reality, the electronic languages that produce this digital domain must be designed and implemented by large international groups of network engineers and computer scientists. Every one of the electronic languages and protocols that make cyberspace possible are carefully designed by specialized international committees. Moreover, once these protocols are established, they only work effectively because the whole network community agrees to abide by these common codes. Without this mutual responsibility, the coherence of cyberspace would quickly break down, because all segments of the Net would no longer be able to communicate with one another. Indeed, they might not be able to communicate at all. The very existence of cyberspace as a globally shared space thus depends on a highly cooperative and mutually responsible community. In this sense cyberspace is a marvelous example of what such a community can achieve—nothing less than the creation of a new global space of being.

There is here an important lesson that I believe we can learn from cyberspace: Any community that shares a "world" is necessarily bound into a *network of responsibility*. Without the continuing support of a community, *any* world (that is, any space of being) will begin to fall apart. If cyberspace teaches us anything, it is that the worlds we conceive (the spaces we "inhabit") are communal projects requiring ongoing communal responsibility.

Some readers will protest at this point. They will object that even if *cyberspace* is a communally produced world, that the physical world is independent of human beings, and that physical space is not reliant on us for its sustenance. On one level this is true: The physical world would not break down if every human being disappeared tomorrow. What *would* break down, however, without continuing communal support is our particular conception of this world—our *worldview*. Consider for example the fundamental shift we have been chronicling in this work, the transition from the medieval worldview (with its dualist conception of spiritual space and physical space) to the modern scientific worldview (with its monistic conception of space). Throughout this transition the physical world *itself* did not change; yet as a matter of lived reality, the world as perceived by the medievals actually *disappeared*. This complex dualistic spatial scheme was replaced by a new monistic spatial scheme with radically different properties. There is a sense in which we must conclude that the medieval world *broke down*—not because the cosmos itself changed character, but because community support for this particular worldview gradually eroded.

Just as cyberspace is communally produced, so in a profound sense are all spaces. Whether we are talking about medieval conceptions of spiritual space, or scientific conceptions of physical space, *every* kind of space must be conceptualized, and hence "produced," by a community of people. Here again, language is key, for every different kind of space requires a different kind of language. Just as cyberspace could not come into being until new kinds of languages for electronic communication had been developed, so *any* new kind of space requires the development of a new language.

Take, for example, astronomical space. In Copernicus' time there simply did not exist a language for talking about cosmological phenomena in physical terms. Over the past four centuries, scientists have gradually developed a sophisticated language of physical cosmology so that those today who study phenomena such as "neutron stars" and "pulsars," the "big bang" and the "Hubble expansion," "gravitational lensing" and "stellar spectrums" can now communicate efficiently and effectively with one another. To name *is*, in a profound sense, to create. And one of the major achievements of the scientific revolution was to articulate a language of physical space. Indeed the creation of new scientific languages is a constant and ongoing part of scientific history. In our own century scientists studying relativistic space have gradually developed *their* own language, as also have those who theorize about hyperspace. Anyone who doubts these disciplines have their own separate languages may like to try reading some of their papers. The fact that each scientific discipline does now have its own language is precisely why it has become so difficult even for other scientists to keep up in fields outside their domain of speciality.

My point here is not to suggest that astronomical space or relativistic space are mere figments of our imaginations, but rather to acknowledge that the "production of space" — any kind of space — is *necessarily* a communal activity. The spaces that we inhabit are irrevocably articulated by communities of people, who cannot express their ideas about reality except through the medium of language. How we see ourselves embedded in a wider spatial scheme is not simply a question of getting to know the "facts"; it is always and ever a matter for social and linguistic negotiation.

As Einstein himself recognized, it is the language we use — the concepts that we articulate and hence the questions that we ask — that determines the kind of space that we are able to see. By shifting the parameters of scientific language, Einstein was able to see a new conception of space. Relativistic space is no fiction (I would not be writing this manuscript at a computer if the designers of microchips had not understood the relativistic effects of electron behavior), but that said, it is important to understand that relativistic space is *not* some "transcendent" reality in the mind of some God. In a very powerful sense it would not exist without Einstein and the subsequent community of relativity physicists. If every relativity physicist died tomorrow and every paper on the subject suddenly disappeared, in what sense could relativistic space be said to "exist"? Just as medieval soul-space disappeared with the demise of the community who supported that concept, so too relativistic space would disappear from the human psychic landscape without the continued sustenance of the physics community.

Since all spaces are necessarily the productions of specific communities, it is not surprising that conceptions of space often reflect the societies from which they spring. Samuel Edgerton has noted that the space of linear perspective was a "visual metaphor" for the orderliness and mercantile rationality of fourteenth-century Florentine society.[30] The anthropologist Emile Durkheim has argued that indeed different societies' conceptions of space always reflect the social organization of their communities. He cites, for example, the Zuni Indians who divide space into seven distinct regions—north, south, east, west, zenith, nadir, and center—which derived from their social experience. According to Durkheim this seven-fold space was "nothing less than the site of the tribe, only indefinitely extended."[31]

As a production of late twentieth-century Western communities, cyberspace, also, reflects the society from which it is springing. As we have noted, this space is coming into being at a time when many in the Western world are tiring of a purely physicalist world picture. Can it be a coincidence that we have invented a new immaterial space at just this point in our history? At just the point when many people are longing once more for some kind of spiritual, or collective psychological, space?

To recognize the contingent nature of our conceptions of space is not to devalue them—relativistic space is no less useful or beautiful because we understand its cultural embeddedness. But in recognizing this, we may become less likely to devalue *other* conceptions of space. The fact that we now live with two very different kinds of space—physical space and cyberspace—might also help us to have a more pluralistic attitude toward space in general. In particular, it might encourage a greater openness toward other societies' spatial schemes. Moreover, if the story we have been tracing in this book has any lesson, I believe it is that our spatial schemes are not only culturally contingent, they are also historically contingent. There is no such thing as an ultimate or supreme vision of space; there is only ever an open-ended process in which we may constantly discover new aspects of this endlessly fascinating phenomenon.

Throughout history new kinds of space have come into being as older ones have disappeared. With each shift in our conception of space also comes a commensurate shift in our conception of our universe—and hence of our own place and role within that universe. In the final analysis, our conception of ourselves is indelibly linked to our conception of space. As I noted at the start of this work, people who see themselves embedded in both physical space and spiritual space cannot help but see *themselves* in a dualistic sense, as physical and spiritual beings. But a people who conceive of space in purely physical terms are virtually compelled to see themselves as purely physical beings. This, of course, is not the only choice; people in non-Western cultures have conceived entirely different options. What *is* universal is that conceptions of space and conceptions of self mirror one another. In a very real sense, we are the products of our spatial schemes.

With the advent of cyberspace we are thus alerted that our conception of our world, and of ourselves, is likely to change. Just as the advent of other kinds of space have always thrown the current worldview into a state of flux, so too cyberspace will likely alter our vision of reality in powerful ways. Just what changes will this new space precipitate? What kinds of reality shifts will it entrain? And how will it affect our conception of our own role within the world system? We cannot yet answer these questions for it is too early to know. In a sense we are in a similar position to Europeans of the sixteenth century who were just becoming aware of the physical space of the stars, a space quite outside their prior conception of reality. Like Copernicus, we are privileged to witness the dawning of a new kind of space. What history will make of this space, appropriately enough, only time will tell.

NOTES

INTRODUCTION

1 Revelations 21:1–24.
2 Marvin Minsky, quoted in Avital Ronell, "A Disappearance of Community." In *Immersed in Technology: Art and Virtual Environments*. Ed. Mary Anne Moser, with Douglas MacLeod. Cambridge, Mass.: MIT Press, 1996, p. 121.
3 Kevin Kelly, quoted from Harper's Magazine Forum "What Are We Doing On-Line?" *Harper's Magazine*, August 1995, p. 39.
4 Michael Heim, "The Erotic Ontology of Cyberspace." In *Cyberspace: First Steps*. Ed. Michael Benedikt. Cambridge Mass.: MIT Press, 1991, p. 61.
5 Michael Benedikt, "Introduction." *Cyberspace: First Steps*, p. 18.
6 Benedikt, ibid., p. 16.
7 Benedkit, ibid., p. 14.
8 Mark Pesce, quoted in Erik Davis, "Osmose." *Wired 4.08*, August 1996.
9 Nicole Stenger, "Mind Is a Leaking Rainbow." In *Cyberspace: First Steps*, p. 52.
10 Hans Moravec, *Mind Children: The Future of Robot and Human Intelligence*. Cambridge, Mass.: Harvard University Press, 1988, p. 124.
11 Umberto Eco, *Travels in Hyperreality*. San Diego: Harcourt Brace Jovanovich, 1986, p. 75.
12 Eco, ibid., p. 75.
13 Gerda Lerner, see *The Creation of Feminist Consciousness: From the Middle Ages to Eighteen-Seventy*. New York: Oxford University Press, 1993. See also Lerner, *The Creation of Patriarchy*. New York: Oxford University Press, 1987.
14 Elaine Pagels, see *The Gnostic Gospels*. New York: Vintage Books, 1989.
15 William Gibson, *Neuromancer*. New York: Ace Books, 1994, p. 6.

16 Quoted in Timothy Druckrey, "Revenge of the Nerds: An Interview with Jaron Lanier." *Afterimage*, May 1991.

17 Moravec, *Mind Children*, p. 4.

18 Allucquere Rosanne Stone, "Will the Real Body Please Stand Up?" In *Cyberspace: First Steps*, p. 107.

19 This report is available online at www.ecommerce.gov.

20 Heim, "The Erotic Ontonology of Cyberspace," p. 73.

21 Quoted in Neil Postman, "Virtual Student, Digital Classroom." *The Nation*, October 9, 1995, p. 377.

22 Heim, "The Erotic Ontology of Cyberspace," p. 61.

23 Nicholas Negroponte, *Being Digital*. New York: Vintage Books, 1996, p. 6.

24 Max Jammer, *Concepts of Space: The History of Theories of Space in Physics*. New York: Dover, 1993, p. 26.

25 Henri Lefevbre, *The Production of Space*. Translated by Donald Nicholason-Smith. Oxford: Blackwell, 1991, p. 2.

26 Sherry Turkle, *Life on the Screen: Identity in the Age of the Internet*. New York: Simon and Schuster, 1995, p. 180.

27 David F. Noble, *The Religion of Technology: The Divinity of Man and the Spirit of Invention*. New York: Alfred A. Knopf, 1997, p. 5.

CHAPTER ONE

1 Dante Alighieri, *The Divine Comedy*. Translated by C. H. Sisson. Oxford: Oxford University Press, 1993. *Paradiso* I: 73.

2 Jacques Le Goff, *The Birth of Purgatory*. Chicago: University of Chicago Press, 1984, p. 290.

3 Giuseppe Mazzotta, "Life of Dante." In *The Cambridge Companion to Dante*. Edited by Rachel Jacoff. Cambridge: Cambridge University Press, 1993, pp. 8–9.

4 John Kleiner, *Mismapping the Underworld: Daring and Error in Dante's Comedy*. Stanford, Calif.: Stanford University Press, 1994.

5 Jeffrey Burton Russell, *Inventing the Flat Earth: Columbus and Modern Historians*. New York: Praeger, 1991.

6 *Inferno* XXXIV:112–114.

7 *Inferno* III:1–3.

8 *Inferno* V:10–12.

9 *Inferno* XXXIV:12.

10 *Inferno* XXXIV:18.

11 John Freccero, "Introduction to *Inferno*." In *The Cambridge Companion to Dante*, p. 175.

12 Freccero, ibid., p. 175.

13 Freccero, ibid., p. 176.
14 Ronald R. MacDonald, *The Burial-Places of Memory: Epic Underworlds in Virgil, Dante, and Milton*. Amherst: The University of Massachusetts Press, 1987, p. 65.
15 *Purgatorio* I:5–6.
16 Le Goff, *The Birth of Purgatory*, p. 339.
17 Jeffrey T. Schnapp, "Introduction to *Purgatorio*." In *The Cambridge Companion to Dante*, p. 195.
18 *Purgatorio* XXI:68.
19 The world of the *Divine Comedy* is full of such subtle reflections and resonances. Indeed Dante scholars through the ages have delighted in finding ever more subtle spatial harmonies throughout this imaginative cosmos.
20 Freccero, "Introduction to *Inferno*," p. 176.
21 Schnapp, "Introduction to *Purgatorio*." In *The Cambridge Companion to Dante*, p. 194.
22 Schnapp, ibid., p. 195.
23 The only exception to this rule are a few Old Testament prophets, who, according to Dante, Christ personally raised up immediately after his death.
24 *Purgatorio* XXXIII:145.
25 *Paradiso* I:136–138.
26 Le Goff, *The Birth of Purgatory*, p. 12.
27 *Purgatorio* XXIII:88.
28 Le Goff, *The Birth of Purgatory*, p. 12.
29 Le Goff, ibid., p. 12.
30 Le Goff, ibid., p. 288.
31 Le Goff, ibid., p. 12.
32 Le Goff, ibid., p. 12.
33 This expression comes from the title of J. Chiffoleau's *La Comptabilité de l'au-delà*. Rome: École Française de Rome, 1980.
34 Revelations 20:12.
35 Le Goff, op. cit., p. 242.
36 *Purgatorio* V:88–136.
37 Rachel Jacoff, " 'Shadowy Prefaces': An Introduction to *Paradiso*." In *The Cambridge Companion to Dante*, p. 215.
38 Jorge Luis Borges, *Seven Nights*. London: Faber and Faber, 1986, p. 6.
39 Le Goff, *The Birth of Purgatory*, p. 177.
40 Le Goff, ibid., p. 32.
41 Jeffrey Burton Russell, *A History of Heaven: The Singing Silence*. Princeton: Princeton University Press, 1997, p. 137.

42 Russell, ibid., p. 137.
43 Russell, ibid., p. 11.
44 Russell, ibid., p. 11.
45 Russell, ibid., p. 11.
46 Russell, ibid., p. 11.
47 Russell, ibid., p. 181.

CHAPTER TWO

 1 John White, *The Birth and Rebirth of Pictorial Space*. London: Faber and
    Faber, 1989, p. 57.
 2 Julia Kristeva, "Giotto's Joy." p. 27.
 3 *Purgatorio*, X:33.
 4 It is interesting to note here that, like Dante's path through Purgatory,
    Giotto's images also follow a right-winding spiral.
 5 White, *The Birth and Rebirth of Pictorial Space*, p. 34.
 6 Hubert Damisch, *The Origin of Perspective*. Translated by John Goodman.
    Cambridge Mass.: MIT Press, 1995, p. xx.
 7 Damisch, ibid., p. 13.
 8 Brian Rotman, *Signifying Nothing: The Semiotics of Zero*. Stanford, Calif.:
    Stanford University Press, 1987, p. 22.
 9 Kristeva, "Giotto's Joy," p. 40.
10 Kristeva, ibid., p. 40.
11 Christine Wertheim, private correspondence with the author.
12 Samuel Y. Edgerton, *The Heritage of Giotto's Geometry: Art and Science on
    the Eve of the Scientific Revolution*. Ithaca, N.Y.: Cornell University Press,
    1991, p. 45.
13 Edgerton, ibid., p. 45.
14 Edgerton, ibid., p. 48.
15 Max Jammer, *Concepts of Space: The History of Theories of Space in
    Physics*. New York: Dover, 1993. See Chapter 3.
16 Quoted in Max Jammer, ibid., p. 11.
17 White, *The Birth and Rebirth of Pictorial Space*, pp. 57–65.
18 Quoted in E. J. Dijksterhuis, *The Mechanization of the World Picture:
    Pythagoras to Newton*. Translated by C. Dikshoorn. Princeton, N.J.:
    Princeton University Press, 1986, p. 162.
19 See here my book *Pythagoras' Trousers*. New York: W. W. Norton, 1997.
    Here I trace the history of the relationship between physics and religion.
20 Jammer, *Concepts of Space*, p. 81.

21 Jammer, ibid., p. 28.

22 Edward Grant, *Much Ado About Nothing: Theories of Space from the* *Middle Ages to the Scientific Revolution.* Cambridge: Cambridge University Press, 1981, p. 100.

23 Note also that Piero's sky is a real physical sky populated with *clouds.* The sheer, flat blue of Giotto has now been replaced with wispy white puffs.

24 Quoted in Samuel Y. Edgerton, Jr., *The Renaissance Rediscovery of Linear Perspective.* New York: Basic Books, 1975, p. 42.

25 Morris Kline, *Mathematical Thought from Ancient to Modern Times,* Vol. 1. New York: Oxford University Press, 1990, p. 233.

26 Rotman, *Signifying Nothing,* p. 19.

27 The whole thing is fake. None of the architectural details are real—it is all painted on a smooth curved ceiling.

28 Michael Kubovy, *The Psychology of Perspective and Renaissance Art.* New York: Cambridge University Press, 1993, pp. 52–64.

29 Rotman, *Signifying Nothing,* pp. 32–44.

30 Edgerton, *The Heritage of Giotto's Geometry.* This is in fact the novel thesis of Edgerton's book, which traces in detail the evolution of perspective and argues for its ultimate significance in the evolution of modern science.

31 Edgerton, ibid., p. 224.

32 E. A. Burtt, *The Metaphysical Foundations of Modern Science.* Atlantic Highlands, N.J.: Humanities Press, 1980, p. 93.

CHAPTER THREE

1 Jeffrey Burton Russell, *A History of Heaven: The Singing Silence.* Princeton, N.J.: Princeton University Press, 1997, p. 126.

2 Russell, ibid., p. XIV.

3 Samuel Y. Edgerton, *The Heritage of Giotto's Geometry: Art and Science on the Eve of the Scientific Revolution.* Ithaca, N.Y.: Cornell University Press, 1991, pp. 195–196.

4 Edgerton, ibid., p. 196.

5 Edgerton, ibid., p. 221.

6 Edgerton, ibid., p. 196.

7 *The New Encyclopaedia Britannica,* vol. 8. Chicago: Encyclopaedia Britannica Inc., 1989, p. 688.

8 Eduard J. Dijksterhuis, *The Mechanization of the World Picture: Pythagoras to Newton.* Princeton, N.J.: Princeton University Press, 1986, p. 226.

9 Alexander Koyre, *From the Closed World to the Infinite Universe.* Baltimore: John Hopkins University Press, 1991, p. 8.

10 Koyre, ibid., p. viii.

11 Jasper Hopkins, *Nicholas of Cusa: On Learned Ignorance.* Minneapolis, Minn.: The Arthur J. Banning Press, 1990, p. 117.

12 Hopkins, ibid., p. 118.

13 Quoted in Koyre, *From the Closed World to the Infinite Universe,* p. 23.

14 Max Jammer, *Concepts of Space: The History of Theories of Space in Physics.* New York: Dover, 1993, p. 84.

15 Hopkins, *Nicholas of Cusa: On Learned Ignorance,* p. 119.

16 Hopkins, ibid., p. 20.

17 Hopkins, ibid., p. 119.

18 Hopkins, ibid., p. 120.

19 Thomas S. Kuhn. *The Copernican Revolution: Planetary Astronomy and the Development of Western Thought.* Cambridge: Harvard University Press, 1985, p. 125.

20 Kuhn. Ibid., p. 125.

21 While the dating of Easter is dependent on the cycles of the sun and moon in Western Christendom, in the Eastern Orthodox Church it has a fixed date.

22 Arthur Koestler, *The Sleepwalkers: A History of Man's Changing Vision of the Universe.* London: Arkana, 1989, p. 143.

23 Fernand Hallyn, *The Poetic Structure of the World: Copernicus and Kepler.* New York: Zone Books, 1990, pp. 94–103.

24 Nicholas Copernicus, *On the Revolutions.* Ed. J. Dobrzycki. Trans. E. Rosen. Baltimore: John Hopkins Press, 1978, p. 4.

25 Those who are interested in this fascinating story can see Chapter Four of my book *Pythagoras' Trousers.* For a more fully detailed account there is Arthur Koestler's marvelous book *The Sleepwalkers,* and for the committed I recommend Fernand Hallyn's magnificent volume *The Poetic Structure of the World.*

26 Owen Gingerich, *The Eye of Heaven: Ptolemy, Copernicus, Kepler.* Washington, D.C.: American Institute of Physics, 1993.

27 Kuhn, *The Copernican Revolution,* p. 155.

28 Koestler, *The Sleepwalkers,* p. 236.

29 Quoted in Koestler, *The Sleepwalkers,* p. 238.

30 Johannes Kepler, *Somnnium: The Dream, or Posthumous Work on Lunar Astronomy.* Trans. Edward Rosen. Madison: University of Wisconsin Press, 1967, p. 28.

31 Galileo Galilei, *Sidereus Nuncius, or The Sidereal Messenger.* Trans. Albert Van Helden. Chicago: Chicago University Press, 1989, p. 40.

32 Giordano Bruno, *The Ash Wednesday Supper*. Ed. and Trans. Edward
Gosselin and Lawrence S. Lerner. Hamden, Conn.: Anchor Books, 1977,
p. 152.
33 Kuhn, *The Copernican Revolution*, p. 132.
34 Bruno, *The Ash Wednesday Supper*, p. 152.
35 Edwin A. Burtt, *The Metaphysical Foundations of Modern Science*. Atlantic
Highlands, N.J.: Humanities Press, 1980, p. 105.
36 Richard S. Westfall, see *Never at Rest: A Biography of Isaac Newton*.
Cambridge: Cambridge University Press, 1990.
37 Quoted in Burtt, *The Metaphysical Foundations of Modern Science*, p. 258.
38 Thomas Hobbes, *The Philosophical Works of Descartes*, vol. 2. Trans.
Elizabeth S. Haldane and G. R. T. Ross. Cambridge: Cambridge
University Press, 1978, p. 65.
39 Burtt, *The Metaphysical Foundations of Modern Science*, p. 104.

CHAPTER FOUR

1 Genesis 1: 14–17.
2 Timothy Ferris, *Coming of Age in the Milky Way*. New York: Anchor
Books, 1989, p. 144.
3 Gale E. Christianson, *Edwin Hubble: Mariner of the Nebulae*. Chicago:
University of Chicago Press, 1995, p. 152.
4 Christianson, ibid., p. 152.
5 Christianson, ibid., p. 143.
6 From the period, one could estimate the *actual* brightness of the Cephid,
then by comparing this with its *apparent* brightness (that is, its appearance
as observed here on the earth), one could calculate its distance away.
7 Robert D. Romanyshyn, *Technology as Symptom and Dream*. New York:
Routledge, 1989, p. 73.
8 Christianson, op. cit., p. 189.
9 Ferris, *Coming of Age in the Milky Way*, p. 208.
10 Ferris, ibid., pp. 209–210.
11 Margaret Wertheim, see Chapter 8, "The Saint Scientific," in *Pythagoras'
Trousers: God, Physics, and the Gender Wars*. New York: W. W. Norton, 1997.
12 Paul Authur Schlipp (Ed.), *Albert Einstein: Philosopher-Scientist*. La Salle,
Ill.: Open Court, 1969. The quote comes from Einstein's
"Autobiographical Notes" in this volume—the nearest he got to ever
writing an autobiography.
13 Letter to Michele Besso, 12 December 1919. In Michele Besso,
Correspondance 1903–1955. Ed. Pierre Speziali. Paris: Hermann, 1972,
pp. 147–149.

14 Gene Dannen, "The Einstein-Szilard Refrigerators." *Scientific American,* January 1997, vol. 276, number 1, pp. 90–95.

15 Max Jammer, *Concepts of Space: The History of Theories of Space in Physics.* New York: Dover, 1993, p. 127.

16 Jammer, ibid., p. 129.

17 Jammer, ibid., p. 141.

18 Quoted in Jammer, ibid., p. 143.

19 Quoted in Banesh Hoffmann, *Relativity and Its Roots.* San Francisco: Freeman, 1983, p. 129.

20 You can try this yourself by taking a balloon and drawing on its surface a collection of spots. As you blow up the balloon, the spots will all move away from each other.

21 Paul Davies, *Superforce: The Search for a Grand Unified Theory of Nature.* London: Counterpoint, 1986, p. 202.

22 Andrei Linde, *Particle Physics and Inflationary Cosmology.* New York: Harwood Academic Publishers, 1990, p. 315.

23 Davies, *Superforce,* p. 202.

24 Lawrence M. Krauss, *The Physics of Star Trek.* New York: HarperCollins, 1995, p. 55.

25 Quoted in Alex Burns, "The Tight Stuff." In *21C: The Magazine of Culture Technology and Science.* Melbourne, Australia: Magazines Unlimited, 2-1997, #23, p. 57.

26 Stephen Hawking, *Black Holes and Baby Universes and Other Essays.* New York: Bantam Books, 1993, p. 116.

27 Hawking, ibid., p. 119.

28 Krauss, *The Physics of Star Trek,* p. 47.

29 Hawking, *Black Holes and Baby Universes and Other Essays,* p. 120.

30 John Gribbin, *Unveiling the Edge of Time: Black Holes, White Holes, Wormholes.* New York: Crown Trade Paperbacks, 1994, p. 175.

31 Andrei Linde, "The Self-Reproducing Inflationary Universe." *Scientific American,* November 1994, vol. 271, number 5, pp. 48–55. See also Lee Smolin, *The Life of the Cosmos.* New York: Oxford University Press, 1997.

32 In fact the *Star Trek* producers appear to have become aware of this problem. In the *Voyager* series the *Enterprise* crew are constantly searching for the "Home Sector." Space itself remains homogeneous and intrinsically undirected, but the producers have given the characters a goal.

CHAPTER FIVE

1 H. G. Wells, *The Time Machine.* London: Everyman, 1993, p. 4.

2 Linda Dalrymple Henderson, *The Fourth Dimension and Non-Euclidean*

Geometry in Modern Art. Princeton, N.J.: Princeton University Press, 1983, p. xix.

3 Henderson, ibid., p. xix.

4 Henderson, ibid., p. 43.

5 J. J. Sylvester, "A Plea for the Mathematician." In *Nature* London. December 30, 1869, p. 238.

6 Edwin A. Abbott, *Flatland: A Romance of Many Dimensions by ASquare.* London: Penguin Books, 1987, pp. 82–83.

7 Michio Kaku, *Hyperspace: A Scientific Odyssey through Parallel Universes, Time Warps, and the Tenth Dimension.* New York: Oxford University Press, 1994, p. 48.

8 Quoted in Henderson, *The Fourth Dimension and Non-Euclidean Geometry in Modern Art,* p. 53.

9 Charles Howard Hinton, *A New Era of Thought.* London: Swan Sonnenschein & Co., 1888, p. 86.

10 Henderson, op. cit., p. 246.

11 Peter Damianovich Ouspensky, *Tertium Organum: The Third Cannon of Thought, A Key to the Enigmas of the World.* (1911) Trans. Claude Bragdon and Nicholas Bessaraboff. New York: Alfred A. Knopf, 1922, p. 327.

12 Henderson, *The Fourth Dimension and Non-Euclidean Geometry in Modern Art,* p. 188.

13 Claude Bragdon, *Projective Ornament.* Rochester, N.Y.: The Manas Press, 1915, p. 11.

14 Quoted in Geoffrey Broadbent, "Why a Black Square?" In *Malevich.* London: Art and Design/Academy Group, 1989, p. 48.

15 Broadbent, "Why a Black Square?" p. 49.

16 Quoted in Henderson, *The Fourth Dimension and Non-Euclidean Geometry in Modern Art,* p. 61.

17 Albert Gleizes, and Jean Metzinger. *Du Cubism.* Trans. and Ed. Robert L. Herbert. *Modern Artists on Art.* Englewood Cliffs, N.J.: Prentice-Hall, 1964, p. 8.

18 Max Jammer, *Concepts of Space: The History of Theories of Space in Physics.* New York: Dover, 1993, p. 152.

19 Quoted in Jammer, ibid., p. 147.

20 Kaku, *Hyperspace,* p. 36.

21 Kaku, ibid., p. 94.

22 Paul Davies, *Superforce: The Search for a Grand Unified Theory of Nature.* London: Unwin Paperbacks, 1986, p. 151.

23 Kaku, *Hyperspace,* p. 103.

24 Quoted in Kaku, ibid., p. 101.

25 Davies, *Superforce*, p. 158.
26 This is so even when quantum effects are taken into account.
27 Davies, *Superforce.* p. 152.
28 Davies, ibid., p. 165.
29 John Gribbin, *Unveiling the Edge of Time.* New York: Crown Publishers, 1992, p. 154.
30 Kaku, *Hyperspace*, p. 98.
31 Robert D. Romanyshyn, *Technology as Symptom and Dream.* New York: Routledge, 1989, p. 43.
32 Romanyshyn, ibid., p. 43.
33 Romanyshyn, ibid., p. 181.
34 Edwin A. Burtt, *The Metaphysical Foundations of Modern Science.* Atlantic Highlands, N.J.: Humanities Press, 1980, p. 95.
35 See here Hubert Damisch, *The Origin of Perspective.* Trans. John Goodman. Stanford, Calif.: Stanford University Press, 1987, p. xix.

CHAPTER SIX

1 Katie Hafner, and Matthew Lyon, *Where Wizards Stay Up Late: The Origins of the Internet.* New York: Simon and Schuster, 1996, pp. 151–155.
2 Hafner, ibid., p. 168.
3 Hafner, ibid., p. 178.
4 Hafner, ibid., p. 242.
5 In the early 1980s bulletin board services (BBSs) also started up, but these were not generally networked together.
6 Howard Rheingold, *The Virtual Community: Homesteading on the Electronic Frontier.* San Francisco: HarperPerennial, 1994, p. 27.
7 Sherry Turkle, *Life on the Screen: Identity in the Age of the Internet.* New York: Simon and Schuster, 1995, p. 180.
8 In fact there is a whole bevy of MUD-type worlds. Other variations are MOOs, MUSHs, MUCKs, and MUSEs. For brevity they are often collectively called MUDs, and that is the term I will use here.
9 Turkle, *Life on the Screen*, p. 12.
10 William Gibson, *Neuromancer.* New York: Ace Books, 1986, p. 51.
11 Rheingold, *The Virtual Community*, p. 156.
12 Turkle, *Life on the Screen*, p. 12.
13 Turkle, ibid., p. 12.
14 Brenda Laurel, see *Computers as Theater.* Addison-Wesley Publishing Company, 1993.
15 Emma Crooker, "Zebra Crossing." *HQ*, Sydney, Australia, July/August 1997, p. 63.

16 Shannon McRae, "Flesh Made Word: Sex, Text, and the Virtual Body." In *Internet Culture*. Ed. David Porter. New York: Rourledge, 1997, p. 79.

17 Turkle, *Life on the Screen*, p. 21.

18 Rheingold, *The Virtual Community*, p. 165.

19 McRae, "Flesh Made Word," p. 79.

20 Turkle, *Life on the Screen*, p. 203.

21 Rheingold, *The Virtual Community*, p. 151.

22 Quoted in Turkle, *Life on the Screen*, p. 10.

23 Barbara Tuchman, A *Distant Mirror: The Calamitous 14th Century*. New York: Ballantine Books, 1987.

24 Erik Davis, "Techgnosis, Magic, Memory, and the Angels of Information." In *Flame Wars: The Discourse of Cyberculture*. Ed. Mark Dery. Durham, N.C.: Duke University Press, 1994, p. 36.

25 John Kleiner, *Mismapping the Underworld: Daring and Error in Dante's Comedy*. Stanford, Calif.: Stanford University Press, 1994, p. 9.

26 Turkle, *Life on the Screen*, p. 14.

27 Turkle, ibid., p. 12.

28 Turkle, ibid., p. 14.

29 Christine Wertheim, Unpublished correspondence with the author.

CHAPTER SEVEN

1 Mark Pesce, "Ignition." Address to "World Movers" conference, January 97, San Francisco. This speech is available online at www.hyperreal.com/~mpesce/.

2 Pesce, ibid.

3 Quoted in Jeff Zaleski, *The Soul of Cyberspace*. San Francisco: HarperEdge, 1997, p. 156.

4 Quoted in by Erik Davis, "Technopagans." *Wired*, July 1995, vol. 3.07.

5 Nicole Stenger, "Mind Is a Leaking Rainbow." In *Cyberspace: First Steps*. Ed. Michael Benedikt. Cambridge, Mass.: MIT Press, 1991, p. 52.

6 Nicole Stenger, ibid., p. 50.

7 William Gibson, *Mona Lisa Overdrive*. New York: Bantam Books, 1988, p. 107.

8 David Thomas, "Old Rituals for New Space." *Cyberspace: First Steps*, p. 41.

9 Mircea Eliade, *The Sacred and the Profane: The Nature of Religion*. San Diego: Harcourt Brace, 1987, p. 23.

10 Stenger, "Mind Is a Leaking Rainbow," p. 55.

11 Eliade, *The Sacred and the Profane*, p. 26.

12 Mary Midgley, *Science as Salvation: A Modern Myth and Its Meaning*. London: Routledge, 1992, p. 15.

13 Michael Benedikt, "Introduction." In *Cyberspace: First Steps*, p. 15.

14 Benedikt, ibid., pp. 15–16.

15 Benedikt, ibid., p. 18.

16 Stenger, "Mind Is a Leaking Rainbow," p. 56.

17 Stenger, ibid., p. 57.

18 Vernor Vinge. *True Names*. New York: Baen Books, 1987, p. 142.

19 Steven Whittaker, "The Safe Abyss: What's Wrong with Virtual Reality?" In *Border/Lines* 33, 1994, p. 45.

20 Caroline Walker Bynam, *Fragmentation and Redemption*. New York: Zone Books, 1992, p. 256.

21 Bynam, ibid., p. 264.

22 Jeffrey Fisher, "The Postmodern Paradiso: Dante, Cyberpunk, and the Technosophy of Cyberspace." In *Internet Culture*. Ed. David Porter. New York: Routledge, 1997, p. 120.

23 Fisher, ibid., p. 121.

24 Vinge, *True Names*, p. 96.

25 N. Katherine Hayles, "The Seduction of Cyberspace." In *Rethinking Tecnologies*. Ed. Verena Andermatt Conley. Minneapolis: University of Minnesota Press, 1993, p. 173.

26 Hans Moravec, *Mind Children: The Future of Robot and Human Intelligence*. Cambridge, Mass.: Harvard University Press, 1988, pp. 109–110.

27 Moravec, ibid., p. 119.

28 Moravec, ibid., p. 122.

29 Rudy Rucker, *Live Robots*. New York: Avon Books, 1994, p. 240. [This volume contains both novels.]

30 AlphaWorld and a variety of other online virtual worlds can be accessed through the Active Worlds website at www.activeworlds.com.

31 Moravec, *Mind Children*, p. 123.

32 Margaret Wertheim, see *Pythagoras' Trousers*. New York: W. W. Norton, 1997.

33 Vinge, *True Names*, p. 112.

34 Thomas, "Old Rituals for New Space," p. 41.

35 Frances Yates, *Giordano Bruno and the Hermetic Tradition*. Chicago: University of Chicago Press, 1979, p. 5.

36 Yates, ibid., p. 32.

37 Yates, ibid., p. 33.

38 Giordano Bruno, "The Expulsion of the Triumphant Beast," quoted in Benjamin Farrington, *The Philosophy of Francis Bacon*. Chicago: University of Chicago Press, 1964, p. 27.

39 Quoted in Francis Yates, *The Rosicrucian Enlightenment*. Boulder, Col.: Shambala Press, 1978, p. 119.

40 David F. Noble, *The Religion of Technology: The Divinity of Man and the Spirit of Invention*. New York: Alfred A. Knopf, 1997, p. 5.
41 Erik Davis, see *Techgnosis: Myth, Magic and Mysticism in the Age of Information*. New York: Harmony Books, 1998.
42 Michael Heim, "The Erotic Ontology of Cyberspace." In *Cyberspace: First Steps*, p. 75.
43 Paulina Borsook, "Cyberselfish." *Mother Jones*, July/August 1996.

CHAPTER EIGHT

1 Avital Ronell. "A Disappearance of Community." In *Immersed in Technology: Art and Virtual Environments*. Ed. Mary Anne Moser, with Douglas MacLeod. Cambridge, Mass.: MIT Press, 1996, p. 119.
2 Howard Rheingold, *The Virtual Community: Homesteading on the Electronic Frontier*. San Francisco: HarperPerennial, 1994, p. 24.
3 Rheingold, ibid., p. 26.
4 Esther Dyson, *Release 2.0: A Design for Living in the Digital Age*. New York: Broadway Books, 1997, p. 2.
5 Dyson, ibid., p. 4.
6 Nicholas Negroponte, *Being Digital*. New York: Vintage Books, 1996, p. 230.
7 Quoted in David F. Noble, *The Religion of Technology: The Divinity of Man and the Spirit of Invention*. New York: Alfred A. Knopf, 1997, p. 40.
8 Quoted in Noble, ibid., p. 38.
9 William J. Mitchell, *City of Bits: Space, Place, and the Infobahn*. Cambridge, Mass.: MIT Press, 1996.
10 Mitchell, ibid., p. 10.
11 Brian A. Connery, "IMHO: Authority and Egalitarian Rhetoric in the Virtual Coffeehouse." In *Internet Culture*. Ed. David Porter. New York: Routledge, 1997, p. 161.
12 Sherry Turkle, *Life on the Screen: Identity in the Age of the Internet*. New York: Simon and Schuster, 1995, p. 240.
13 Turkle, ibid., p. 239.
14 Connery, "IMHO: Authority and Egalitarian Rhetoric in the Virtual Coffeehouse," p. 166.
15 Connery, ibid., p. 175.
16 Connery, ibid., p. 176.
17 Michele Tepper, "Usenet Communities and the Cultural Politics of Information." In *Internet Culture*, p. 43.
18 Connery, "IMHO: Authority and Egalitarian Rhetoric in the Virtual Coffeehouse," p. 176.

19 Stephanie Brail, see "The Price of Admission: Harassment and Free Speech in the Wild, Wild West." In *Wired Women: Gender and New Realities in Cyberspace*. Eds. Lynn Cherny and Elizabeth Reba Weise. Seattle, Wash.: Seal Press, 1996.

20 Brail, ibid., p. 148.

21 Brail, ibid., p. 152.

22 Progress and Freedom Foundation, "Cyberspace and the American Dream: A Magna Carta for the Knowledge Age." This document is available online at www.pff.org

23 Quoted in Mary Fuller and Henry Jenkins, "Nintendo and the New World Travel Writing: A Dialogue." In Steven G. Jones (Ed.), *Cybersociety: Computer-Mediated Communication and Community*. London: Sage, 1995, p. 59.

24 Ziauddin Sardar, "alt.civilization.faq: Cyberspace as the Darker Side of the West." In *Cyberfutures: Culture and Politics on the Information Superhighway*. Ed. Ziauddin Sardar and Jerome R. Ravetz. New York: New York University Press, 1996, p. 19.

25 Sardar, ibid., p. 19.

26 This simulation was called SIMNET. See Lev Manovich, "The Aesthetics of Virtual Worlds." In *CTHEORY*, vol. 19, no. 1–2. This document is available online at www.ctheory.com.

27 Sardar, "alt.civilization.faq: Cyberspace as the Darker Side of the West," p. 22.

28 For information about online racist hate groups, see "163 and Counting . . . Hate Groups Find Home on the Net." In *Intelligence Report*, Winter 1998, vol. 89. Montgomery, Ala.: Southern Poverty Law Center, 1998, pp. 24–28.

29 See Henri Lefebvre, *The Production of Space*. Trans. Donald Nicholson-Smith. Oxford: Blackwell, 1991.

30 Samuel Y. Edgerton Jr., *The Renaissance Rediscovery of Linear Perspective*. New York: Basic Books, 1975, pp. 30–40.

31 Quoted in Stephen Kern, *The Culture of Time and Space 1880–1918*. Cambridge, Mass.: Harvard University Press, 1983, p. 138. [See Emile Durkheim and Marcel Mauss, *Primitive Classification*. 1970.]

INDEX

Page numbers in *italics* refer to illustrations.